T0318460

South Africa's Energy Transition

South Africa's energy transition has become a highly topical, emotive and politically contentious topic. Taking a systems perspective, this book offers an evidence-based roadmap for such a transition and debunks many of the myths raised about the risks of a renewable-energy-led electricity mix. Owing to its formidable solar and wind resources, South Africa has an almost unparalleled opportunity to turn solar photovoltaic and onshore wind generators into the country's power generation workhorses – a role hitherto played by coal. This book shows that a renewables-led mix will not only provide the lowest cost, but will also create more jobs than any of the alternatives currently under consideration. In addition, it offers a glimpse of how South Africa's low-cost and decarbonised electricity system can power a competitive industrial economy, an electric-mobility revolution and, in the long run, create new export opportunities.

This book will be of great interest to energy industry practitioners, as well as students and scholars of energy policy and politics, environmental economics and sustainable development.

Tobias Bischof-Niemz is the Head of Corporate Business Development at ENERTRAG AG, a Berlin-based global clean-energy company. He previously established and led the Energy Centre at the South African Council for Scientific and Industrial Research (CSIR) in Pretoria, and holds a PhD from TU Darmstadt, Germany, and an MPA from Columbia University, New York, USA.

Terence Creamer is a South African journalist who has been writing about the country's energy sector since 1994. He is currently editor of Creamer Media's *Engineering News* and a contributing editor to *Mining Weekly* and *Polity*.

Routledge Focus on Environment and Sustainability

The Application of Science in Environmental Impact Assessment
Aaron Mackinnon, Peter Duinker and Tony Walker

Jainism and Environmental Philosophy
Karma and the Web of Life
Aidan Rankin

Social Sustainability, Climate Resilience and Community-Based Urban Development
What About the People?
Cathy Baldwin and Robin King

South Africa's Energy Transition
A Roadmap to a Decarbonised, Low-cost and Job-rich Future
Tobias Bischof-Niemz and Terence Creamer

The Environmental Sustainable Development Goals in Bangladesh
Edited by Samiya Selim, Shantanu Kumar, Rumana Sultana and Carolyn Roberts

For more information about this series, please visit: www.routledge.com/ Routledge-Focus-on-Environment-and-Sustainability/book-series/RFES

"We started the Energiewende in 2000 amidst uncertainties: Which technologies will win? How much state support before they reach cost competitiveness? But we moved ahead regardless, and were successful. Today, new solar parks and wind generators are the cheapest forms of electricity generation – by far; and South Africa can reap the benefits."

—Rainer Baake, *former Minister of Energy, Federal Republic of Germany*

"This book explores options at hand for progressive governments to move towards a diversified energy mix that stimulates inclusive economic growth and increases energy access, and responsibly manage the transition to a lower-carbon economy. These alternative energy technologies are fast approaching, like a tsunami, and require governments to be agile."

—Dr Tsakani Lotten Mthombeni, *Chairperson, Energy Intensive Users Group of Southern Africa (EIUG), South Africa*

"While the German Energy Transition is moving into the second phase, South Africa can reap the benefits of being a 'follower country' that does not have to pay the – sometimes painful – school fees of early adopters anymore, but can move with full speed into the least-cost, least-carbon energy future."

—Jörg Müller, *Founder and CEO, ENERTRAG AG, Germany*

"Drawing on their considerable insight into South Africa's technical power supply system and its political economy, the authors present riveting reflections on energy transition milestones and illustrate in practical, even-handed terms how it could accelerate national development and regain South Africa's global status as a preferred electricity investment destination."

—Brenda Martin, *CEO, South African Wind Energy Association (SAWEA), South Africa*

South Africa's Energy Transition

A Roadmap to a Decarbonised, Low-cost and Job-rich Future

Tobias Bischof-Niemz and
Terence Creamer

Routledge
Taylor & Francis Group
LONDON AND NEW YORK

earthscan
from Routledge

First published 2019
by Routledge

2 Park Square, Milton Park, Abingdon, Oxfordshire OX14 4RN
52 Vanderbilt Avenue, New York, NY 10017

Routledge is an imprint of the Taylor & Francis Group, an informa business

First issued in paperback 2019

British Library Cataloguing-in-Publication Data
A catalogue record for this book is available from the British Library

Library of Congress Cataloging-in-Publication Data
Names: Bischof-Niemz, Tobias, author. | Creamer, Terence, author.
Title: South Africa's energy transition : a roadmap to a decarbonised, low-cost and job-rich future / Tobias Bischof-Niemz and Terence Creamer.
Description: Abingdon, Oxon ; New York, NY : Routledge, 2019. | Series: Routledge focus on environment and sustainability
Identifiers: LCCN 2018019534 | ISBN 9781138615168 (hardback) | ISBN 9780429463303 (ebook)
Subjects: LCSH: Energy policy—Environmental aspects—South Africa. | Renewable energy sources—Economic aspects—South Africa.
Classification: LCC HD9502.S56 B57 2019 | DDC 333.790968—dc23
LC record available at https://lccn.loc.gov/2018019534

ISBN: 978-1-138-61516-8 (hbk)
ISBN: 978-0-367-34010-0 (pbk)

Typeset in Goudy
by Apex CoVantage, LLC

Contents

Foreword

I have learned many valuable lessons while promoting the advancement of science, technology and innovation in South Africa over the past two decades. However, three lessons stand out. Firstly, scientists can never rest on their laurels, with past achievements ever vulnerable to the next technological breakthroughs. Secondly, in a context of massive societal backlogs and scarce resources, trade-offs are required. In making those difficult choices, the main determinant should always be the delivery of the highest developmental return. The third lesson is that, while good science is vital, so too is sound policy and political commitment.

All three lessons, I believe, are equally germane to South Africa's energy milieu and to the technological transition that is beginning to take shape, principally in the electricity sector.

South Africa cannot rest on its past energy glories, particularly in light of the fact that those achievements have been built largely on solutions that are dirty, thirsty and harmful to human and environmental health. Technological breakthroughs have already taken place in the electricity environment and others, both known and currently unknown, are undoubtedly on the way. Tough political decisions are now required to ensure that electricity remains affordable and reliable and that it is not only more accessible, but also produced in a way that is environmentally sustainable.

The good news contained within the pages that follow is that South Africa has a genuine opportunity to deliver on all these goals by progressively tapping into its world-class solar and wind resources. Such a transition may take some time to evolve, but is both technically feasible and commercially viable. Backed by a strong political vision and the correct policy choices, this renewable-energy-led electricity system can deliver an industry that is not only cheaper, cleaner and less water intensive than the alternatives, but which is also more jobs rich.

This is a massive opportunity that should be embraced, promoted and put into action.

Dr Sibusiso Sibisi
Director and Head of Wits Business School

Preface

With the contrail from air force jets still visible in the cloudless April sky over Pretoria's Union Buildings, President Khanyisile Moya takes to the podium to address the nation. Only days earlier, Moya's party had emerged victorious following the hotly contested 2049 electoral race. Political commentators are putting the victory down to her party's message of continuity and change, which captured the national mood. As ever, South Africans are hungry for improvements in education, healthcare and security. Moya, an electrical engineer, was particularly effective in offering a compelling vision for a refashioned educational system, closely aligned to the changing demands of the country's knowledge-based industrial economy. Business has become increasingly vocal in its call for employees to be equipped with far deeper analytical capacity so they are able to make sense of the information riches flowing from factories, farms, mines, offices and households.

For continuity, Moya did not have to look far for inspiration. Having read for an electrical-engineering degree after leaving high school in 2010, she became a renewable-energy pioneer and ended up owning and managing one of the first wholly black-woman-owned clean energy companies, as they were then known in the early 2020s. Before selling her shares and entering into politics, Moya was often described as the most influential energy entrepreneur in Africa and a leading global figure in a sector that, during her career, underwent one of the most dramatic transitions of any industry in history. However, she always attributed her success and that of her company to South Africa's energy policy, which remains the envy of the world.

The country's energy framework was established some 30 years ago when Moya's visionary predecessors recognised that South Africa had the unique opportunity to convert its unrivalled solar and wind resources into the lowest-cost and cleanest electricity in the world. The decision had been highly contested at the time, owing to vested interests in the coal and nuclear industries. However, policymakers were alive to the steep fall in solar photovoltaic (PV) and wind technology costs and seized the moment.

The decision sparked a manufacturing revival, built initially around a large-scale domestic renewables roll-out, but then increasingly on the export of green technologies into Africa and the rest of the world. Later, it also led to a revival in energy-intensive manufacturing as South Africa's power prices became a global competitive advantage. Indeed, the country's entire industrial policy began to revolve around an electrification-of-almost-everything vision, which saw the country emerge not only as an international hub for electric vehicle production, but also for low-carbon fuels, chemicals, fertilisers and a plethora of metal and engineering products. The visionary policy decision had not only transformed Moya's life, but the lives of all South Africans.

The narrative here is precisely the 2050 vision of South Africa's energy system that this book explores, explains and advocates.

Acknowledgments

We wish to thank all those who made it possible for us to write this book. Professionally, we have gained invaluable insight and inspiration from many talented and supportive colleagues at Creamer Media, ENERTRAG, the Council for Scientific and Industrial Research and Eskom. We are also grateful for the access we have had during our respective careers to some of South Africa's most influential energy professionals and policymakers.

As both of us have full-time jobs, the decision to research and write this book had a direct impact on our families, who have offered nothing but support. In fact, Valérie Bischof-Niemz, Tobias' wife, was the initial inspiration behind the project and remained a firm believer in the book's merits even when the going got tough. Debbie French, Terence's wife, didn't only provide irreplaceable moral support, but also played an invaluable role in proofreading the early drafts. Tobias' sons – Noah, Léo and Mylan – as well as Terence's sons – Andrew and Mark – are our main motivation for thinking about a better energy future. They were encouraging and understanding throughout.

We would also like to express our gratitude to the team at the Taylor & Francis Group for their professional management of this project, as well as for their thoughtful and diligent advice and guidance.

Tobias Bischof-Niemz & Terence Creamer

Introduction

Cheap, abundant and unconstrained

'By using energy generated from our abundant supplies of coal to produce aluminium, the smelter will contribute to the drive to add more value to our primary products before export', President Nelson Mandela said in his address to dignitaries at the opening of the Alusaf Hillside smelter in Richards Bay, KwaZulu-Natal, on 19 April 1996.[1] The speech was made less than two years after South Africa's first democratic elections on 27 April 1994 – an historic event that marked an end to decades of brutal colonial and apartheid rule and the start of a new era of political liberation.

Achieving economic emancipation, however, remained a remote aspiration. The country's economy – undermined by years of race-based exclusion, militarisation, Bantu education, international sanctions, protectionist policies as well as serious political and fiscal mismanagement – remained fragile. In fact, only days before Mandela's Alusaf address, attended overwhelmingly by the white businessmen who still occupied just about all positions of authority in the private sector, the markets had reacted fiercely to the appointment of the country's first black Finance Minister, Trevor Manuel. Also the first member of Mandela's African National Congress to occupy the position, Manuel succeeded banker Chris Liebenberg,[2] who was perceived by the markets as a safe pair of hands. Prior to Liebenberg, the role was filled by Derek Keys, who many largely credited for the decision to purchase and invest in Alusaf[3] while leading mining and industrial conglomerate Gencor, and who also served as Finance Minister in the Cabinet of FW De Klerk; South Africa's last head of state from the era of white-minority rule. The Rand, which has long been a proxy for market sentiment towards South African political developments, lost 10% of its value in two months,[4] breaching the R4 to the dollar level. And while growth was picking up and inflation declining, there was little room in the Budget for tax, or on-Budget incentives to bolster investment. In fact, in his final Budget speech,[5] Liebenberg acknowledged that the 'general pattern of the last few years has been to limit the number of tax incentives', and to focus instead on 'support measures in respect of training, investment, exports, preferential market access, technology, work organisation, productivity and small business development'.

However, government did have one lever at its disposal for stimulating fixed investment: cheap and abundant electricity. South Africa's power surplus had arisen as a result of a consistent overestimation of demand growth for the 1980s and beyond[6] by planners at the Electricity Supply Commission (Escom), now Eskom. As a consequence, between 1970 and 1994, the vertically integrated utility added 33 910 MW of new generation capacity. Along the way, tariffs surged and there was even a period of load shedding, owing to a lag between investment decision-making and plant commissioning. After shock price increases in the late 1970s of up to 30% and 45% a year in nominal terms, another period of sustained hikes of between 15% and 23% followed in the 1980s.[7] The second round of increases provoked a national outcry, which resulted in then President PW Botha appointing the De Villiers Commission of Inquiry into 'The Supply of Electricity in the Republic of South Africa'. The commission's recommendations led to changes in the Electricity Act in 1985 and to new Eskom and Electricity Acts in 1987.[8] The actions taken following the De Villiers Commission improved the financial and commercial performance of Eskom, supported by an initial increase in tariffs, a reduction in the pace of capital expenditure, as well as Eskom's tax- and dividend-exempt status.[9] Eskom also entered into a pricing compact with the government, whereby it committed itself to a price decrease of 20% between 1992 and 1996, and a 15% reduction between 1994 and 2000. Actual price reductions were a little less than this.[10]

The period of the Mandela administration coincided with a recovery in Eskom's reputation, as tariff reductions erased memories[11] of the hikes that occurred in the 1970s and 1980s. It could even be argued that it was an era of cheap, abundant and unconstrained power, with the international debate on the harmful effects of coal-fired electricity generation on the climate only starting to gather momentum. During the period, government actively sought to secure energy-intensive investments. Besides the Alusaf investment, which was followed by a further potline expansion of the Hillside smelter, in KwaZulu-Natal, Gencor's successor BHP Billiton pursued the Mozal smelter in Maputo, Mozambique, again on the back of Eskom's cheap power offer. Again the tariff was linked to the aluminium price, offering Hillside and Mozal downside protection during periods of price weakness and Eskom some upside during periods of aluminium strength. So overwhelming was perception of electricity abundance that, in the early 2000s, South Africa was still actively pursuing negotiations, initially with Pechiney and later with Pechiney's acquirer Alcan, for another proposed smelter investment for the

Coega industrial development zone in the Eastern Cape. The project was eventually abandoned in 2009, when Rio Tinto, which acquired Alcan, pulled the plug as the extent of South Africa's power shortages became apparent. At the same time, BHP Billiton's special pricing agreement with Eskom was coming under heavy criticism amid a rise in load shedding. The contracts remained, but the terms of the Mozal contract were revised in 2010, with the link to the aluminium price abandoned.

What this period of perceived abundance disguised was that low prices did not necessarily equate to efficient performance.[12] South Africa lost sight of the consequences of Eskom's massive overinvestment, as well as the fact that good investment decisions have larger impacts on electricity prices than any post-build incremental productivity improvements. In the event, South Africa was destined to repeat the mistakes of the 1970s and 1980s when it started running short of electricity from 2007 onwards – an event ironically signalled in the Department of Energy's (DoE's) own 1998 White Paper Energy Policy. The paper stated that 'although growth in electricity demand is only projected to exceed generation capacity by approximately the year 2007, long capacity-expansion lead times require strategies to be in place in the mid-term, in order to meet the needs of the growing economy'[13]. The response of government and Eskom to the shortfall was to initiate a large, coal-based build programme, funded by steep tariff hikes, together with debt and some direct fiscal support. It was a response that all but ignored Eskom's previous capital-investment inefficiencies, eschewed calls for institutional reform and dismissed changes under way internationally relating to generation technology costs and the structure of the electricity supply industries. Eskom's inflexible construction programme has since resulted in a significant and growing surplus of expensive generation capacity.[14] In the process, demand has been destroyed as energy-intensive businesses either shut capacity, intensified energy-efficiency drives, or invested in cheaper supply alternatives. Eskom's Medium-term System Adequacy Outlook, published in July 2017, forecasts an expected excess capacity of just over 8 GW in 2022 based on a low-demand scenario. This has led to a discussion about the possibility of accelerating the decommissioning of the Grootvlei, Hendrina and Komati power stations and cancelling the construction of units five and six at the 4.8 GW Kusile coal-fired project being built in Mpumalanga.

Adding to the problem has been the political contamination of electricity planning processes. Decision-making in electricity is highly politicised[15] and has also been characterised by a lack of transparency and by power struggles. A battle over which technologies should be prioritised has been a key feature, along with concerted efforts to ensure that nuclear is included in any future mix. The politicisation of electricity decision-making has led to delays in the updating of the country's Integrated Resource Plan (IRP), which, at the time of writing, had not been updated since 2011. A joint paper by Amory Lovins and Anton Eberhard, titled 'South Africa's Electricity Choice', highlights the level of political meddling during President Jacob Zuma's tenure as president. Following the unsuccessful attempts to update the IRP in 2013 and 2016, the DoE approached Eskom's modellers during the last quarter of 2017 to produce another update to reflect lower actual electricity demand and the latest comparative energy prices.[16] The new electricity demand forecasts were still too high, as were the cost assumptions for renewables. By contrast, the assumptions for nuclear costs were too low. 'But even with these conservative assumptions, nuclear energy was not picked in any of these new South African modelling scenarios, other than one where artificial constraints are placed on how much solar and wind energy can be built and where additional carbon budget limits are imposed. Even in this extreme scenario, nuclear energy would only be required after 2039', the paper notes. These outcomes were not welcomed by the nuclear section in the DoE and the modellers were asked to run a scenario where 9.6 GW of nuclear power stations are 'hardwired' or 'forced' into the model.[17]

Such planning distortions have been commonplace. The Draft IRP Base Case of 2016 was widely criticised as having not represented the least-cost planning outcome, as the document included artificial[18] limitations on the amount of solar photovoltaic (1 000 MW) and onshore wind (1 600 MW) that could be added to the system yearly. Since Zuma's resignation and the swearing in of Cyril Ramaphosa as President on 15 February 2018, there is an expectation that electricity planning might become less politicised and less distracted by efforts to ensure that the nuclear build programme proceeds. Speaking on the sidelines of the World Economic Forum in Davos, Switzerland, only weeks before becoming president, Ramaphosa said: 'We have to look at where the economy is – we have excess power and we have no money to go for a major nuclear plant building'.[19]

Nevertheless, the political-economy factors are likely to remain critical in determining the direction that South Africa's electricity sector takes in the coming years and decades. This book fully recognises this reality. However, its focus is primarily on exploring the techno-economic prospects for transitioning South Africa's coal-dominant electricity system to one that is built on South Africa's solar and wind resources. In other words, the intention of the book is to provide the knowledge base on which a meaningful policy discussion can be held. By doing so, the book will seek to prove that South Africa can return to an era of cheap, abundant and, critically, environmentally unconstrained electricity.

This book is premised on the argument that the South African system, while in transition, has been 'disrupted'. This may appear to be a contradictory statement, given that a transition implies a prolonged period of adjustment, while disruption implies a sudden change. However, the words 'disruption' and 'transition' are both applicable when discussing development in South Africa's electricity supply industry. The fact that there has been a disruptive change in the cost of generation technologies is now indisputable. Five years ago, it would have been cost-optimal to build new coal-fired power stations, or even new nuclear in a context where South Africa was required to reduce its carbon emissions. However, owing to the sharp fall in the cost of onshore wind and solar photovoltaic (PV) generation, a tipping point has been reached, whereby it no longer makes economic sense to build new coal, or new nuclear, in South Africa. The consequence of that disruptive change is immediate for decision-making; as the new-build decisions shift from new coal or nuclear, with some wind and solar PV, to a system dominated by wind and solar PV, and supported by flexible generation. Yet, this 'disruption' will lead only to a gradual change in South Africa's actual supply mix. Put another way, the immediate disruption in decision-making will still only result in a 'transition' from a coal-led system to one based on wind, solar and flexible generation. The long lifetime of the existing asset base means that the change is gradual. In other words, a digital switch in new-build decisions today will lead to significant changes in the supply mix only 10 to 20 years down the road. However, it will lead to that with mathematical precision. The effects of this transition make the disruptive change quite manageable.

To explore the techno-economic aspects of the South African transition, this book relies on well-established power system engineering logic for designing a least-cost power system. Such a systems engineering approach presupposes an interdisciplinary process of ensuring the design and implementation of a high quality, trustworthy, cost-efficient solution across the system's entire life cycle. In sum, systems engineering is an interdisciplinary approach and means to enable the realisation of successful systems.[20] In this book such an approach is premised on ensuring that as many of the cheapest kilowatt-hours possible are harnessed; that any gaps are filled by more expensive, but highly responsive kilowatt-hours and that the system's overall risks are mitigated.

Hitherto, the cheapest generation option for South Africa has been coal. However, as discussed previously and elaborated upon in detail in the chapters that follow, this is no longer the case. The steep decline in the cost of solar PV and onshore wind has made these generation sources cheaper than either coal or nuclear, even after accounting for the costs associated with filling the supply gaps that arise when the sun goes down or the wind stops blowing. This tipping point is the product of years of developed-country subsidisation, which follower countries can now avoid. In the South African context, where the combined solar and wind resource is world class, if not best in class, that cost competitiveness is further amplified. This book demonstrates that it is both technically feasible and economically rational to turn these resources into the workhorses of the country's electricity system.

In other words, designing the entire future power system around solar and wind. The opportunity does not end there, however. South Africa can also progressively decarbonise its entire energy system by coupling its low-cost, renewables-led electricity generation to the transportation and heating sectors. At the same time it could create new industrial opportunities and even become the equivalent of a Saudi Arabia for clean chemicals and fuels.

That's not to say there are no risks associated with a transition to a higher penetration of variable renewable energy. There is a greater need for flexibility to respond to this variability and, once penetration levels breach certain thresholds, there is also a need for technology and operational solutions to guarantee stability. However, there are no showstoppers, with most of these solutions already available. What's more, these will not add materially to overall system costs. A risk is also posed by South Africa's vertically integrated utility model, which is not suitable for the transition. Indeed, Eskom is already showing signs of severe distress, as costs rise, supply falls and tariffs surge. A new model, which separates Eskom's generation assets from its transmission and system operation units, is likely to prove more resilient to the disruption already under way, as well as to any future shocks to the functioning of the electricity supply industry. However, by fully embracing the transition, there is significant upside potential for South Africa. In a context where countries are progressively moving to decarbonise their energy systems through investments into wind and solar, South Africa's superior natural resources will make its electricity comparatively cheaper. By making the correct policy choices, South Africa could reposition itself as the investment destination of choice for any activity that is electricity intensive.

The energy transition is, thus, an opportunity for South Africa to recapture the competitive advantage referred to by President Nelson Mandela in his address at the opening of the Alusaf aluminium smelter. This time, though, the country's cheap and abundant electricity supply will not face environmental constraints. However, reaping these benefits will require clear-sighted policymaking, bold decision-making and determined leadership. This leadership may need to draw inspiration and courage from a quote so often attributed, rightly or wrongly, to South Africa's great statesman: 'It always seems impossible – until it's done'!

Notes

1 www.sahistory.org.za/archive. Speech by President Nelson Mandela at the opening of the Alusaf Hillside Smelter Richards Bay, 19 April 1996.
2 Davis, G., Roussouw, Rehana and Wackernagel, Madeleine. *Mail & Guardian*, Liebenberg replaced by Manuel, 29 March 1996. https://mg.co.za/article/1996-03-29-liebenberg-replaced-by-manuel/.

3 Gencor Ltd. *Company Profile, Information, Business Description, History, Background Information on Gencor Ltd.* www.referenceforbusiness.com/history2/22/Gencor-Ltd.html#ixzz59uuMQSDr.
4 Pippa Green. *Independent Online. The outsider who has measured vision against reality.* 16 February 2006. www.iol.co.za/business-report/economy/the-outsider-who-has-measured-vision-against-reality-744408.
5 www.gov.za. Budget Speech 1996 Minister of Finance Mr CF Liebenberg, 13 March 1996.
6 Steyn, G. *Investment and Uncertainty: Historical Experience with Power Sector Investment in South Africa and Its Implications for Current Challenges,* 15 March 2006.
7 Steyn, G. *Administered Prices. Electricity – A Report for National Treasury,* 2003.
8 Eberhard, A. *The Political Economy of Power Sector Reform in South Africa,* 2009. Publishing by Cambridge University Press in *The Political Economy of Power Sector Reform: The Experiences of Five Major Developing Countries.* Edited by Victor, D. and Heller, T.
9 Steyn, G. *Investment and Uncertainty: Historical Experience with Power Sector Investment in South Africa and Its Implications for Current Challenges,* 15 March 2006.
10 Eberhard, A. *The Political Economy of Power Sector Reform in South Africa,* 2009. Publishing by Cambridge University Press in *The Political Economy of Power Sector Reform: The Experiences of Five Major Developing Countries.* Edited by Victor, D. and Heller, T.
11 Eberhard, A. *The Political Economy of Power Sector Reform in South Africa,* 2009. Publishing by Cambridge University Press in *The Political Economy of Power Sector Reform: The Experiences of Five Major Developing Countries.* Edited by Victor, D. and Heller, T.
12 Eberhard, A. *The Political Economy of Power Sector Reform in South Africa,* 2009. Publishing by Cambridge University Press in *The Political Economy of Power Sector Reform: The Experiences of Five Major Developing Countries.* Edited by Victor, D. and Heller, T.
13 Department of Energy, *White Paper on the Energy Policy of the Republic of South Africa,* December 1998.
14 Steyn, G., Burton, J. and Steenkamp, M. *Meridian Economics: Eskom's Financial Crisis and the Viability of Coal-Fired Power in South Africa,* November 2017.
15 Baker, L., Burton, J., Godinho, C. and Trollip, H. *University of Cape Town Energy Research Centre. The Political Economy of Decarbonisation: Exploring the Dynamics of South Africa's Electricity Sector,* November 15.
16 Lovins, A. and Eberhard, A. 'South Africa's Electricity Choice', January 2018.
17 Lovins, A. and Eberhard, A. 'South Africa's Electricity Choice', January 2018.
18 *Ministerial Advisory Council on Energy.* Comments on the Integrated Resource Plan 2016 Draft South African Integrated Resource Plan 2016 public hearing. Presentation made in Johannesburg, 7 December 2016.
19 Dahinten, J. and Burkhardt, P. for Bloomberg News, *Ramaphosa Says South Africa Has No Cash for Nuclear Plants,* 25 January 2018.
20 *International Council on Systems Engineering.* www.incose.org/AboutSE/WhatIsSE.

1   A tipping point

During the height of South Africa's load-shedding crisis in 2015, Robbie van Heerden, Eskom's system operator at the time, would leave home before dawn and make the short journey from his home in the south of Johannesburg to Eskom National Control, which lies about 20 km east of the city. He wanted to be positioned in front the control room's multi-storey video wall well before 5 am, when the country's electricity demand would begin its inevitable climb towards its morning peak.

Situated alongside the Victoria Lake yacht club, in Germiston, and only a few kilometres down the road from the world's largest precious metals refinery, the Rand Refinery, the white-tile-clad building does little to betray its status as a national key point. Nevertheless, the National Control was thrust unceremoniously into the limelight in 2015 as citizens learned just how important it was in determining whether daily rotational power cuts would take place and for how long. South Africa was gripped, at the time, with the threat of a 'national black-out', a recovery from which could take more than two weeks. Hourly load-shedding radio reports became essential listening, with collective sighs of relief when cuts were restricted to 'Stage 1', or 1 000 MW of rotational cuts. Anxiety levels rose considerably, though, on those days when 'Stage 3' (3 000 MW) load shedding was communicated. Such announcements generally set social media ablaze with warnings of the kind of catastrophic black-out load shedding was, ironically, designed to prevent.

Apart from the reputational damage suffered by Eskom, the power cuts came with some very real social and economic consequences, losing about a percentage point of gross domestic product and lowering business confidence. For a businessperson, the cuts had serious top- and bottom-line implications, affecting operating hours, as well as staffing, maintenance and investment decisions. Energy-intensive companies bore the brunt of it. They were not only forced to trim consumption by 10%, but were also first in line for additional curtailment, before the unpopular decision of load-shedding households was taken. For private citizens, the schedules determined everything from homework and leisure routines to when and how meals were prepared, with a tiny minority of callers to radio talk shows actually claiming to enjoy the candle-lit ambiance delivered courtesy of Eskom. For the system operators at National Control there was no place for levity, however. Every day, but especially during the morning and evening peaks, was a white-knuckle ride, with unplanned plant outages making system forecasting and scheduling almost impossible.

In the control room, the full-colour screens glowed with numbers and graphs; no matter how dark the surrounding areas might have been at the time. For his part, Van Heerden would focus almost exclusively on one number and one screen. That screen showed system frequency as measured in Hertz (Hz), while the number he was looking for was always 50. Eskom, as is the case with most system operators globally, works within a frequency band of between 49.5 Hz and 50.5 Hz. The goal, though, is always 50 Hz, which is indicative of a controlled flow of alternating current power from multiple generators through the network, or the sum of the output of multiple generators instantaneously equalling the sum of all customer demand. Generation output is changed every four seconds to match customer demand. But when there is insufficient generation available, as was frequently the case during the first half of 2015, customer demand is reduced to ensure system stability. Van Heerden uses the analogy of a fully-laden truck moving up a hill at 30 km/h. If the engine capacity is too small for the truck's load, the engine slows, which in electricity terms, translates to a drop in system frequency. To prevent the truck engine from stalling, load is shed or, in the case of electricity, controlled manual load shedding is initiated.

On a winter morning in South Africa, demand can rise from as low as 24 000 MW at 5 am to over 30 000 MW by 7 am, where it plateaus for most of the working day. Demand begins to rise again at around 5 pm, picking up by a material 3 000 MW in about an hour, before peaking just after 6 pm. Every day is different, though, and during the 2015 load-shedding crisis, even small supply or demand events had the potential to destabilise a system being held together in the absence of the normal 2 000 MW operating reserve. National Control was, thus, continually on the hunt for additional supply or ways to reduce demand. Arrows in the quiver included the diesel-fuelled open-cycle gas turbines, demand-reduction contracts with industry, some imports and the Hillside aluminium pot lines, which could be switched off for a few hours on a weekly rotation. There was also tremendous vigilance for any 'peaks within the peak', which could result in the frequency falling precipitously. Every day provided a new adrenalin rush, Van Heerden reflects. However, those daily anxieties would have never arisen had the authorities heeded National Control's warning, made in the late 1990s, that new capacity should be in place by 2007 to guarantee security of supply.

This nightmare scenario for South Africans, which played itself out most intensively in 2015, but which emerged as a real threat all the way back in 2006, is only partly attributable to a failure of planning. In the event, the growth-sapping gap between theoretical capacity and actual plant availability arose as result of a series of calamitous policy, regulatory and corporate decisions. The Department of Energy's own 1998 White Paper Energy Policy stated that 'although growth in electricity demand is only projected to exceed generation capacity by approximately the year 2007, long capacity-expansion lead times require strategies to be in place in the mid-term, in order to meet the needs of the growing economy'.[1] Nevertheless, the events brought home the importance of planning, as well as adherence to plans. Indeed, had Eskom implemented, or been allowed to implement, its own plan, the first new capacity would most likely have been ramping up at the very time when the power cuts were becoming a daily threat.

System planning

Credible power system planning begins with a demand forecast, as well as with assumptions about the residual lifetime of existing generation assets. In any power system, the supply gap widens over time, either because demand is increasing, or because an existing plant is being decommissioned, or both. In other words, a supply gap arises not only because of increases in demand, but also because, at some point, old power stations become too expensive and inefficient to keep operational.

The supply gap needs to be filled with new generation assets, and ideally it needs to be filled a few years ahead of time so as to prevent the type of load-shedding described previously. This is as true for centrally planned power systems, such as South Africa's, as it is in more liberalised markets. The only difference being that, in a centrally planned system, the demand/supply forecast is done by the central planner, while in liberalised power systems the individual market participants are forecasting demand and supply-side options independently. The overall objective is the same: to determine if and when a new-build power station investment decision makes sense. Settling on such a forecast is an intricate process. It involves the use of models that produce a forecast using input assumptions relating to economic growth, as well as the power-intensity of that growth.

Forecasts will always diverge from reality, which is why system planners habitually produce more than one scenario and rely on regular updating to keep deviations in check. In the case of South Africa, forecasters typically calculate low-, medium- and high-growth scenarios, with the demand forecast eventually selected for use by power-system planners arising from a consultation process involving system operators at Eskom, academics, large customers and research councils. The forecast selected for the 2010 version of South Africa's Integrated Resource Plan (IRP) proved wildly optimistic. The document, which was promulgated in 2011, predicted that electricity demand would grow by 2.8% a year between 2010 and 2030,[2] which would have seen yearly electricity demand double over the period to 454 TWh, from 250 TWh. This demand includes not only South African end-customer consumption, but also exports and grid losses.

In reality, though, domestic electricity consumption has flattened since 2010 and the Energy Intensive Users Group (EIUG), which represents the 32 large companies that consume about 40% of South Africa's electricity, believes demand assumptions should be moderated even further.[3] The organisation says South Africa's electricity demand has not grown since 2007, owing to several factors related to the change in the economic structure of South Africa, including the lack of generation capacity.

The forecast has since been moderated considerably (see Figure 1.1), reflecting an assumed growth path that rises from 245 TWh a year to only around 382 TWh a year in 2050. Undoubtedly, this demand assumption will also be proved incorrect. Nevertheless, in light of South Africa's stage of development, as well as the country's ongoing population growth and its industrialisation ambitions, planners believe it prudent to plan for expansion, as underestimating demand will prove more damaging than any overestimation. Assuming lower demand would be to expect South Africa's already relatively modest per-capita consumption to fall even further, which would arguably imply economic catastrophe. Indeed, even if South Africa achieved the IRP 2016's original yearly demand assumption of 522 TWh by 2050 (currently deemed exceedingly optimistic) the country would simply have increased per-capita consumption to a level currently enjoyed in Australia – a mining economy with many structural similarities. It would also discount prospects for new demand, which could arise from electric vehicles, or increasing shares of electrification in industrial and residential heat provision. If South Africa has learned anything from its recent load-shedding crisis surely it must be that the negative consequence of stranded generation assets is significantly smaller than not having sufficient electricity.

On the supply side, meanwhile, any phasing out of existing generation assets is also forecast to create a complete picture of the anticipated generation gap. In the case of South Africa, which has a relatively mature fleet of coal power stations, the retirement of plants, particularly from around 2030 onwards, cannot be ignored (see Figure 1.2).

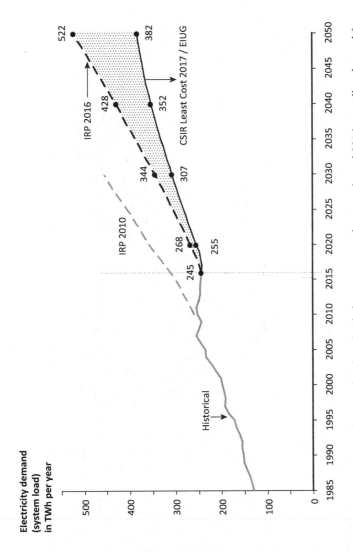

Figure 1.1 Historic electricity demand in South Africa (including exports to the region) until 2016, as well as demand forecasts from IRP 2010, IRP 2016 and CSIR Least Cost 2017 / EIUG until 2050

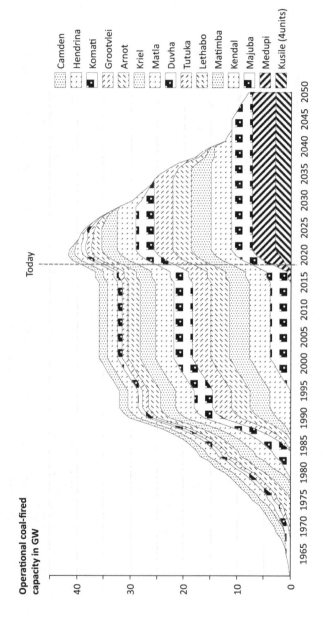

Operational coal-fired capacity in GW

Today

Camden
Hendrina
Komati
Grootvlei
Arnot
Kriel
Matla
Duvha
Tutuka
Lethabo
Matimba
Kendal
Majuba
Medupi
Kusile (4units)

40 — 30 — 20 — 10 — 0

1965 1970 1975 1980 1985 1990 1995 2000 2005 2010 2015 2020 2025 2030 2035 2040 2045 2050

Figure 1.2 Existing Eskom coal fleet and its scheduled decommissioning dates over time

Once the demand forecast and the decommissioning schedule of the existing power stations are agreed, system planners then begin the task of matching new-build supply options to the widening supply gap, using a mathematical model that has but one objective function: determining the least-cost generation mix over the planning horizon within a firm, non-negotiable boundary condition of stipulated system reliability. Besides cost, contemporary planners also strive for an outcome that ensures a reduced environmental impact. Therefore, additional boundary conditions are included in the model, outlining the specific environmental thresholds that should not be breached, such as carbon-emission limits.

Planners then go about filling the supply gap in the least-cost manner, subject to the constraints imposed. They determine the least-cost mix by inputting individual cost items associated with the available generation technologies, such as the capital outlay to build the plant, the fixed operations and maintenance cost to own the plant, and the variable cost (predominately fuel) to operate the plant, together with the technical specifications of the different supply options. The mathematical model then determines how – with all the supply options available, their costs and their technical characteristics – the future electricity demand can be met at the least cost, while remaining within the boundary conditions of system reliability and environmental constraints. The objective function and the key output of such a model, together with the cost-optimal investment mix, is the present value of the total cost to build and operate the power system from today until the end of the planning horizon, say 2050.

Levelised cost of energy (LCOE)

Another output is the predicted utilisation of the individual plants that the model has selected to build. The utilisation, together with the cost per technology, leads to the cost per energy unit produced. In the case of South Africa, the cost is stated in Rand per kilowatt-hour (R/kWh) for each technology. This is called the Levelised Cost of Energy, or LCOE. It is important to note that one can only calculate LCOE by making an assumption about what the utilisation of a certain power plant will be. The utilisation of the different power stations is an output, though, of a system-optimisation model, and not an input. LCOE of different technologies is nevertheless an important first indicator for planners (see Figure 1.4).

Using the analogy of a motor vehicle, the R/kWh value of a power station would be equivalent to the Rand per kilometre (R/km) travelled by the vehicle. For the individual car owner the R/km includes both fixed and variable costs. The fixed costs are associated with capital-related costs, including loan repayments, and fixed operations and maintenance costs, such as insurance, service plans, time-related maintenance. These are incurred regardless of use. The variable costs, meanwhile, relate to fuel, as well as the distance-related wear and tear. These costs scale with utilisation. The aim when buying a vehicle is to minimise the R/km travelled. To achieve this, the buyer's specific demand profile must be taken into account. If the vehicle is going to be driven only for 5 000 km a year, its fuel consumption is of far less importance than if the vehicle is being used for daily commuting, resulting in 30 000 km of yearly travel. In the latter case, even if the car is more expensive to buy in the first place (capital outlay), the buyer may still be willing to make the investment to secure a better fuel consumption.

In the case of a power plant, there are also fixed and variable costs. The variable costs are mostly fuel-related and are incurred only when the plant is producing electricity. The fixed costs are again associated with capital-related costs (i.e. the cost of building the power station), as well as with fixed operations and maintenance cost (e.g. the staff costs at the power station, the insurance cover, the scheduled time-related maintenance). Stated differently, all fixed costs are triggered once the plant is built and need to be paid for regardless of whether the plant is used or not, while variable costs are only triggered when the power station is in use.

The proportion of fixed to variable costs in the LCOE is technology dependent. Capital-intensive nuclear power stations have a higher proportion of fixed costs in their LCOE when compared with coal plants, where variable cost items, such as coal, limestone and water, comprise about one-third of the total cost. Gas-fired plants are predominantly weighted towards variable costs, because they are relatively cheap to build and maintain, but the fuel is expensive (at least in South Africa where natural gas fuel is generally much more expensive than coal fuel per energy unit). Interestingly, the most capital-intensive power stations of all are solar photovoltaic installations and wind farms, because they have no variable costs whatsoever.[4]

The LCOE is the sum of all three cost components (see Figure 1.3): the annualised capital expenditure divided by the energy produced in a year, the fixed operation and maintenance costs divided by the energy produced in a year and the fuel cost associated with producing one additional unit of energy. The numerator of the first two components is fixed, which means the cost per energy unit is higher if less energy is produced. The third component, being fuel, is a variable cost, which will only arise when energy is produced. Per unit, a variable cost component is therefore constant.

As with a car that is not used often, the unit cost, in R/km, goes up, the same applies for power generators. If they are used less often, the unit cost in R/kWh, or the LCOE, rises. How quickly the LCOE increases, however, depends on the capital intensiveness of the technology. The more capital intensive a technology, the faster the LCOE increases with reduced utilisation.

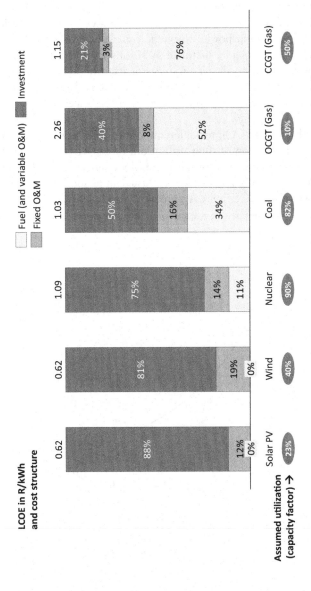

LCOE in R/kWh and cost structure

Legend: ■ Fuel (and variable O&M) ■ Investment ■ Fixed O&M

	Solar PV	Wind	Nuclear	Coal	OCGT (Gas)	CCGT (Gas)
LCOE in R/kWh	0.62	0.62	1.09	1.03	2.26	1.15
Investment	88%	81%	75%	50%	40%	21%
Fixed O&M	12%	19%	14%	16%	8%	3%
Fuel (and variable O&M)	0%	0%	11%	34%	52%	76%
Assumed utilization (capacity factor) →	23%	40%	90%	82%	10%	50%

Figure 1.3 Cost structure of different power-generation technologies, sorted from capital-intensive, to the left to capital-light, or fuel-intensive, to the far right; OCGT (Gas) = Open-Cycle Gas Turbine, fired with natural gas; CCGT (Gas) = Combined-Cycle Gas Turbine, fired with natural gas, all numbers in April-2016 prices

For solar PV and wind, the utilisation is determined by the weather, whereas for all other power generators, utilisation depends on the operator's dispatch decision. For power stations the term used to describe the 'utilisation' is capacity factor. The capacity factor describes what the average output of the power generator is during a certain time period, relative to the maximum power output it can deliver as a result of its installed capacity. A capacity factor of 30% for example does not mean that the power generator always produces at exactly 30% of its capacity, nor does it imply that it produces at full capacity 30% of the time. It means that, on average, the power generator produces at 30% of its nominal installed capacity.

Figure 1.4 shows the LCOE for different conventional power generation technologies as a function of their utilisation, or capacity factor. For the most capital-intensive conventional technology, nuclear, the LCOE is relatively modest at a high utilisation of 90%. But as soon as the utilisation decreases, the LCOE increases quickly. At roughly 70% capacity factor, nuclear is already more expensive than a gas-fired power station

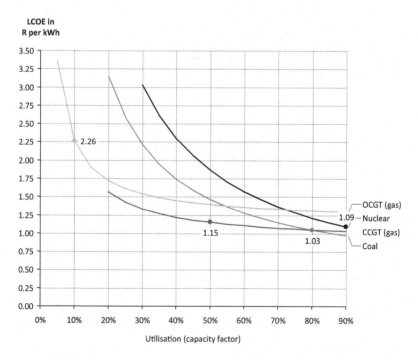

Figure 1.4 How LCOE changes with different capacity factors for different technologies

with an open-cycle gas turbine (OCGT). The OCGT is capital-light and fuel-intensive, which means that, with further decreasing capacity factor, its cost advantage over nuclear grows, because the low utilisation does not hurt a capital-light power plant's LCOE as much as it does a capital-intensive one. If a power station with a 55% capacity factor or less is required, then the gas-fired OCGT is even cheaper than a coal-fired power station.

The choice of the appropriate technology for a certain purpose depends heavily on the expected utilisation. As with the car example, if the expected utilisation is low – perhaps because the power station fulfils a back-up function, or is supposed to supply customer demand only in the daytime – the choice tends to be more towards capital-light power stations that are cheap to build and maintain, but burn expensive fuel.

System-wide levelised cost of energy

In South Africa, a simulation software known as PLEXOS[5] is used to determine the LCOE not only for a single power station, but for the entire fleet of power stations. The solution outlines the overall LCOE not only of today's power stations, but that of today's and all future new-build power stations over the planning horizon, of say 30 to 40 years. In other words, it is used to minimise the system LCOE/total system cost per energy unit over the planning time horizon, in present value terms. This is not the same as minimising the LCOE of each and every power station in the system. The model does this by taking into account all the interdependencies between the different power generators.

Using the motor vehicle analogy, it is equivalent to determining the least-cost vehicle mix for a large logistics company. The objective is not to minimise the R/km cost of each individual vehicle, but to minimise the overall cost of fleet operations. To achieve such optimisation, there will be some vehicles that are expensive to buy, but cheap to operate (i.e. low fuel consumption), which will be used as the 'workhorses' of the fleet. However, there will also be those vehicles that are cheap to buy and are used as back-up in those cases where the main workhorses in the fleet fail, or to make short-trip emergency deliveries. And then there are those vehicles in between, which are only used at daytime and only weekdays, which means with a half-loading of potential kilometres they could technically travel, the fuel cost of those are not as important as for the 'workhorses', but more important than for the 'emergency' vehicles.

In the power system, the modelling process is used to determine the cost-optimal mix of power generators to guide future investment decisions.

In other words, the plan, or in the case of South Africa, the IRP, offers guidance on what technologies to install, how much of each of them and by when. Once the least-cost, or base case, mix has been determined, planners then run various scenarios to determine the cost effects of any deviation from the base case. The scenarios could cater, for instance, for a policy decision to 'decarbonise' the electricity system over a defined time horizon, or to introduce constraints on certain technologies. The idea is to offer visibility of the cost implications of various policy choices. It also provides policymakers with a rational techno-economic basis for their decisions and allows citizens to reach their own value-for-money conclusions.

In answering the fundamental question as to what to build next, planners adopt a long-term perspective so as to try and avoid building assets that could be stranded in future. The optimal pathway forward is a least-cost, least-regret balance. Generally, planners like granular investments with short lead and lifetimes, and low risk of time and cost overruns. That makes the entire planning process more agile and less prone to mistakes that have consequences reaching far into the – unknown – future. The risk of stranded assets has grown in recent years as a result of the steep reduction in some technology costs, most notably the cost of wind and solar photovoltaic, combined with the increasing global availability of natural gas through the shipping of that fuel as liquefied natural gas (LNG) across the world. Planners globally are being forced to adjust their models to these new realities, fully conscious of the fact that another technological breakthrough is more than likely to happen during the planning time horizon. The planning 'sweet spot' for accommodating such 'unknown unknowns' is when the lowest-cost solution is also the lowest risk, i.e. when the lowest-cost technologies also have the shortest lead times and lifetimes. Such a convergence has not always been possible.

Historically, the first place South African planners turned in order to meet the objective function of minimising the cost of the supply was coal. The combination of the country's abundant coal resources, together with the fact that the environmental costs were not internalised, meant that the fixed and variable costs associated with coal-fired power stations, or their LCOE, remained well below that of all alternatives.

This implied that, for decades, the cheapest power generators in terms of LCOE (coal-fired power stations) were also the ones with the longest lead times. Coal-fired power station projects were large and carried a significant risk of cost and time overruns. Planners had to accept these risks, because the mathematical objective function of a 'least-cost system LCOE' demanded that this cheap, coal-based electricity forms a significant part of the mix – and the risks associated to it had to be accepted. There was no similarly cheap alternative.

Furthermore, owing to the large role that mining and other power-intensive industries play in the South African economy, the country has a relatively high load factor, or baseload demand. Of the 198 TWh a year of final domestic consumption recorded in South Africa with its 55 million inhabitants, International Energy Agency (IEA) statistics show that industry accounts for 122 TWh, or 62%. Residential demand is 37 TWh, or 19%, while commercial and public services account for 14% or 27 TWh. This profile is likely to become more 'peaky' in future as commercial and residential demand grows and more citizens move into the middle class. For instance, in Australia, with a population of 24 million and where total consumption is 211 TWh, the IEA statistics show that industry comprises only 77 TWh, or 36%. Residential consumption is 59 TWh, or 28%, while commercial and public services is 67 TWh, or 32% (see Table 1.1).

Under South African conditions of high base demand and baseload supply being the cheapest available technology on a power-plant-LCOE level, it made sense for planners to meet demand primarily through the building of large coal-fired power stations. In addition, nuclear was added in an effort to lower the risk associated with transmitting electricity form the north-eastern parts of the country, primarily the Mpumalanga province, to the key load centre in the Western Cape, more than 1 500 km to the southwest. Both technologies are capital intensive and, therefore, if there is a choice, the objective is always to use the plants as much as possible. This was relatively simple under the historical paradigm, as the system operator determined utilisation. The scenario is changing, though, with the introduction of renewable energy, where production of electricity from solar and wind power plants cannot be controlled and planned in the same manner as before. Nevertheless, until

Table 1.1 Composition of electricity demand in South Africa versus Australia[6]

Final electricity demand in TWh/a	Australia (2015)	South Africa (2015)
Others	8 (4%)	12 (6%)
Residential	59 (28%)	37 (19%)
Commercial	67 (32%)	27 (14%)
Industry	77 (36%)	122 (62%)
Total	**211 (100%)**	**198 (100%)**
Population	24 million	55 million

Note: Percentages may not add up to 100%, as they are rounded to the nearest percent.

recently it was only logical to deploy plants that were expensive to build, but cheap to operate, at high rates of utilisation.

As a consequence, South African electricity planners – and planners around the globe, for that matter – employed a 'bottom-up' model to address the supply/demand gap. They turned first to the most capital-intensive solution, nuclear, and on top of that, added layers of coal-fired stations, which could be more economically throttled during periods of low demand, owing to their higher variable costs. The coal plants where added, again from the bottom, based on a merit order, whereby the plants with the lowest LCOE at high utilisation were introduced first.

At some level of supply, though, a point is reached where it no longer makes economic sense to meet the residual demand with baseload power stations, because that residual demand is not baseload, 24/7 demand, but rather a daytime or weekday demand. At this point, supply solutions with lower capital intensiveness, which are cheaper in LCOE terms at lower capacity factors – such as hydro pumped storage, gas or diesel – are deployed to meet the variable customer demand profile that emerges over and above the constant load factor, derived primarily from mining and industrial demand. The additional power stations deployed to meet the residual demand will have a lower capacity factor. Finally, so-called peaking plants are introduced for unforeseen circumstances and to supply the peak demand. The role of the peaking stations is akin to the logistics fleet manager referred to earlier who buys cheap vehicles to fulfil irregular, unexpected or peak-demand deliveries, being less concerned about their fuel consumption than about fulfilling orders. Likewise, medium fuel-efficient vehicles will be deployed for daily daytime tasks, while highly fuel-economical, expensive-to-buy trucks will be used for the jobs that require almost 24/7 utilisation.

When base generators, such as coal and nuclear, were the cheapest in terms of individual power station LCOE, system planners turned to these solutions first (see Figure 1.5). However, this was not because such generators perfectly matched the load, because they did not. Customer demand, and hence the electrical load in the power system, is changing constantly during the day, the week and over the different seasons. It is by no means constant or 'baseload'. Nevertheless, the aim was to push as much of the cheapest kilowatt-hours into the system as possible and only turn to more expensive, flexible technologies when it became uneconomical to insert yet another rectangular chunk of new generation into the system. Strict mathematics, following a system-LCOE minimisation logic, determined that coal and nuclear were the workhorses, providing the most kilowatt-hours to the total system supply.

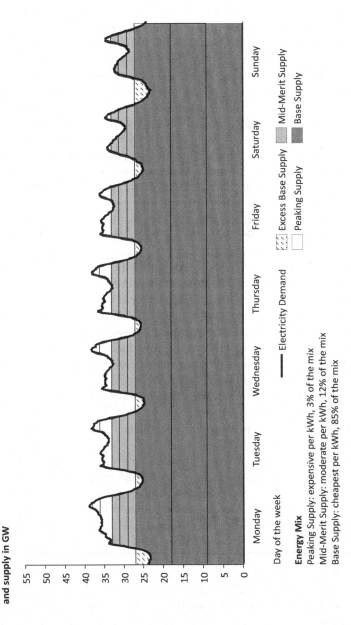

Electricity demand and supply in GW

Day of the week

— Electricity Demand

🔲 Excess Base Supply 　 ▨ Mid-Merit Supply

⬜ Peaking Supply 　 ⬛ Base Supply

Energy Mix
Peaking Supply: expensive per kWh, 3% of the mix
Mid-Merit Supply: moderate per kWh, 12% of the mix
Base Supply: cheapest per kWh, 85% of the mix

Figure 1.5 Stack of supply options that power-system planners used to meet the demand

Logic remains, outcome changes

In the new world, the design logic discussed previously remains intact. Neither the laws of physics, nor the mathematical tools, nor the objective function of a least-cost system LCOE have changed. The difference is that the cheapest kilowatt-hour on an individual-technology-LCOE basis is now produced by wind and solar PV, and no longer by baseload coal. And that cost advantage is not by a small margin, but already today in South Africa by roughly 40%, with the gap widening as wind and solar PV become cheaper year on year. This, it should be stressed, is a new phenomenon, as, until recently, wind and solar PV were not the lowest-cost technology choices and where, therefore, only fringe contributors to the overall power system. However, the dramatic fall in the cost of supplying power from wind and solar PV plants has taken the industry beyond a critical 'tipping point'; irreversibly altering planning outcomes and disrupting the traditional operating model.

With the cheapest kilowatt-hours no longer generated by baseload plants, or better termed 'base supply' plants, planners are no longer populating supply/demand models from the bottom. Because the cheapest electricity now comes from a technology that is being dispatched by the weather, not by human choice, the entire model is turned on its head (see Figure 1.6). In this new world, wind and solar PV transitions from playing only a modest and supportive role to becoming the basis of power supply, with the rest of the power system being optimised around them.[7] That's not to say system planners and operators no longer place a premium on dispatchability, they do. But economics dictates that the lowest-cost generation sources be deployed first, even if these are variable and do not match the customers' demand directly by themselves. As with the base-supply-first model, residual demand in a renewable-first world is met by using more expensive, but flexible, kilowatt-hours.

In other words, the planning paradigm still seeks to push as much of the cheapest kWh into the system as possible. Simply because they are the cheapest. That might even lead to oversupply situations where the workhorses (either base-supply power generators or solar PV/wind) produce more of the cheap kWh than the customers demand. This happens to coal and nuclear power stations, as much as is the case with solar PV and wind. The economics of such curtailed 'excess energy' need to be factored into the plan. Any excess kWh increases the effective cost of the useful kWh from the workhorse electricity supplier. The only significant difference between the two planning paradigms is that in the solar PV and wind world this excess energy is relatively more than in the base-supply world. As long as solar PV and wind are cheap enough compared to, say, new coal, the economics work.

**Electricity demand
and supply in GW**

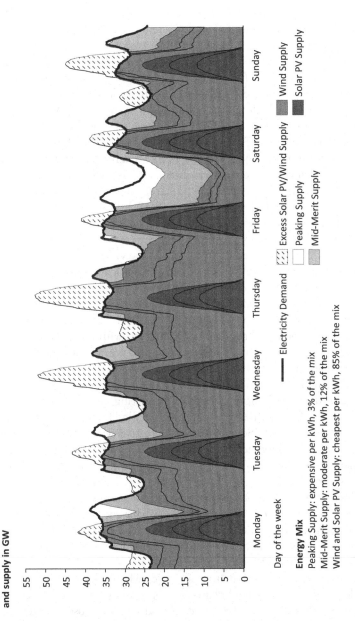

Figure 1.6 Stack of supply options in a cost-optimal mix with cheap solar PV and wind

Technically feasible, economically viable

But is such a mix feasible? Until recently, there has been a view that a transition to a renewables-led power system would require an entire overhaul of the power system. However, there is growing evidence to suggest that only an evolution of the current system is required. In fact, there is a growing list of countries that are already making the transition. To operate a power system on hydropower is – if available – technically trivial and not new. But hydropower is limited in its potential. What is more interesting for the transition are high penetration levels with so called variable renewables (VRE), i.e. solar PV and wind, which were at 0% in most countries not long ago. Today, there are a number of countries that are already running their power systems on double-digit VRE penetration levels, such as Denmark (55%), Uruguay (30%), Germany (27%), Ireland (26%), Portugal (24%), Spain (22%), United Kingdom (20%), Greece (19%), Honduras (18%) and Nicaragua (16%).[8]

Over the last decade or so, the focus of power-system planners has clearly shifted from 'technical feasibility' to 'economic viability'. Where technical feasibility describes whether something is technically possible within the laws of physics, while economic viability asks whether something is the cheapest or at least an affordable thing to do. Without any doubt, a high-renewables power system is technically feasible. The economic viability is breaching tipping points country by country, with those with the highest solar and wind resources and the least area constraints, such as South Africa, being first to be beyond the tipping point.[9]

In South Africa, a renewable-energy-led mix has been stress tested using a simulated time-synchronous model,[10] integrating wind and solar data from the Wind Atlas South Africa and the Solar Radiation Data respectively. The outcome reflects South Africa's impressive wind and solar resource base, with a capacity factor of wind turbines of 35%, and with solar PV capacity factors of 20 to 25% found to be achievable almost anywhere in the country.

One of the implications of this dramatic change is that the cost-optimal mix for a country such as South Africa, which has such extraordinary renewables resources, is now also the most environmentally desirable – carbon emissions fall, along with water use. In other words, there is no longer any trade off necessary between clean electricity and least-cost electricity.[11]

However, the transition from an order where electricity is supplied predominantly by baseload plants that operate around the clock to one where weather-dispatched renewables lead the way, poses challenges. If left unmanaged, these challenges can threaten system reliability, as well as the sustainability of markets designed around vertically integrated utilities. For instance, the investment case even for flexible fossil-fuel power stations is likely to come into question, owing to the fact that the plants operating hours are likely to fall below viability levels without changes to the current market settings. By 2022, Germany is forecasting that total load will be completely covered by renewable energy alone for many hours throughout the year,[12] which means that the back-up fossil fuel plants could remain out of service for protracted periods. For this reason, regulatory remedies may be required to incentivise investments that support system reliability.

Nevertheless, the dramatic decline in the cost of wind and solar PV is an undeniable reality and will not only change the way power markets are planned and operated, but will also have far-reaching environmental and social consequences. The starting point, however, has to be an acknowledgement that wind and solar PV have now passed a cost-competitiveness tipping point. Once that is accepted, planning becomes far less complicated. One possible advantage that South Africa has in being able to adopt a system approach to the development of its electricity industry lies in the policy weight given to the country's IRP. Once the plan is adopted, the Energy Minister is able to allocate capacity across the selected technologies, using determinations that are formally published in the *Government Gazette*. If a technology is absent in the plan, it cannot be procured or built, making the IRP both a technology roadmap and an investment blueprint. In other words, it acts as a powerful planning and policy instrument.

As discussed earlier, if implemented as theoretically envisaged, the IRP could offer a high degree of investor certainty. This is because the document provides good visibility of the expected evolution of the generation mix over a multi-decade horizon, while directing the ministerial determinations that contain the shorter-term investment commitments. For South Africa to achieve the status of a preferred electricity investment destination will not, therefore, require any major policy overhaul. It will, however, require improvements to the way the IRP is drafted, as well as a far more disciplined approach to implementation. And it will require that investment decisions are made and implemented even if they affect the incumbent Eskom as a vertically integrated utility adversely.

In light of the long-term implications of the plan, the Department of Energy should seek to safeguard the initial drafting process from political contamination. This can be achieved by contracting technical specialists to oversee the compilation of a least-cost base case, derived using fully transparent parameters and input assumptions, as well as modelling mechanics. In other words, it should be a techno-economic exercise, unsullied by politics or policy. This base case should be subjected to public scrutiny and will provide a long-term picture to 2050. If policy adjustments are to be made, these should be made after the fact, with the rationale and cost premium caused by any deviation made public.

The entire process should be fully transparent and no model input, output or mechanics (the physics) should be introduced or used without also being made available to stakeholders. Such an approach could go a long way to mitigate the potential for political manipulating of the plan. It should also help close trust deficits that could arise around the drafting process and any policy adjustments that follow. It would also follow international best practice.[13]

Continuous improvement

To ensure that the IRP remains relevant and credible, though, South Africa's energy planners should commit to publishing and promulgating regular updates. In addition, once the plan is approved, the first five to eight years should be locked in through the publication of a formal investment plan (see Figure 1.7), which provides investors with a high degree of certainty regarding the country's technology allocations, as well as execution dates. The IRP should then be updated and optimised yearly, but the five- to eight-year investment plan should remain hardwired on a rolling basis. With every IRP update, an additional year would be added to the investment pipeline. Should there be a fall in demand, this could be reflected in either there being no new investment allocations for that year, or a low level of new commitments. By coupling planning transparency with execution diligence, the investment climate would be de-risked to the point where South Africa could secure generation projects on a highly cost-competitive basis.

Such an approach of continuous checks and balances with regard to the planning assumptions will help avoid stranded assets or undersupply. To use such an approach of continuous adjustment is feasible today, because the least-cost capacity expansion plan for South Africa no longer includes any new coal nor nuclear. These are the only long-lead-time and high-timeline-risk assets in the portfolio of available power generators. With them not being an economic option anymore, planning becomes easier, because individual investment decisions become smaller, more granular, with shorter lead times and less risk of time overruns. These are the properties of the technology choices in the 'new electricity world', where solar photovoltaic, wind power and smaller gas-fired power stations come to the fore.

Bottom Line: The fundamental system-planning logic for a least-cost power system remains: Including as much of the cheapest kilowatt-hours into the system as possible, while filling the gaps with more expensive generation. What has changed is that new wind and solar PV, not coal, are now the cheapest generation sources by far, and should, thus, be the workhorses of South Africa's future electricity system.

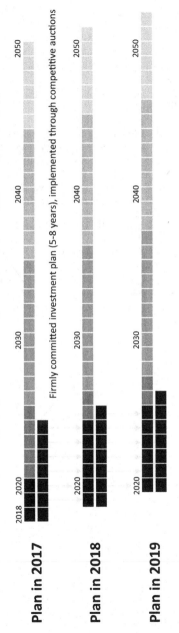

Figure 1.7 Proposal for linking of long-term IRP with a firmly committed short-term investment plan. If such a plan was adopted in 2017 (the starting point for such a framework), the first three years of the new plan until 2020 would be hard coded to include renewables assets already procured, but not yet implemented

Notes

1 Department of Energy. *White Paper on the Energy Policy of the Republic of South Africa*, December 1998.
2 Department of Energy. *Integrated Resource Plan 2010–2030*, 2011.
3 Energy Intensive Users Group. *EIUG Comment on the IRP2016*, 31 March 2017.
4 CSIR Energy Centre. presentation on 'Electricity Sector Expansion Planning in South Africa', March 2017.
5 Energy Exemplar Website, accessed November 2017, https://energyexemplar.com/software/plexos-desktop-edition/.
6 *International Energy Agency Statistics*. www.iea.org.
7 Agora Energiewende. '12 Insights on Germany's Energiewende', February 2013.
8 REN21, Renewables 2018 Global Status Report, page 43, www.ren21.net/wp-content/uploads/2018/06/17-8652_GSR2018_FullReport_web_-1.pdf.
9 Brown, T.W., Bischof-Niemz, T., Blok, K., Breyer, C., Lund, H., Mathiesen, B.V., Response to 'Burden of proof: A comprehensive review of the feasibility of 100% renewable-electricity systems', *Renewable and Sustainable Energy Reviews*, https://doi.org/10.1016/j.rser.2018.04.113.
10 Creamer, T., Engineering News, 'CSIR Model Makes Case for a Renewables-First Electricity Mix', September 2016.
11 Creamer, T., Engineering News, 'Electricity-Mix Model Shows South Africa Won't Need to Trade Off Clean for Low-Cost', February 2017.
12 Agora Energiewende, '12 Insights on Germany's Energiewende', February 2013.
13 Pfenninger, S., *Energy scientists must show their workings*, Nature 542, 393 (23 February 2017), doi:10.1038/542393a, https://www.nature.com/news/energy-scientists-must-show-their-workings-1.21517.

2 How did we get here?

Convalescing after knee surgery at the university hospital in Mainz, a city on the west bank of the Rhine river, a bored Matthias Willenbacher found himself paging through the *Mainzer Allgemeine Zeitung* newspaper. Deep inside its regional section, the young physics doctoral student was drawn to an article about the development of a wind turbine near Prüm, another Rhineland-Palatinate town about 200 km to the west. The developers of the turbine boasted that it could generate enough electricity to supply 200 households. Growing up in a farming family, Willenbacher decided early on in life that agriculture was not for him. Nevertheless, he felt immediately inspired by this new type of farming: wind harvesting. 'If the people of Prüm could do it, I could do it', was Willenbacher's immediate reaction, narrated in his 2014 book, *My Indecent Proposal to the German Chancellor*.[1]

Despite initial resistance from his father and obstacles thrown up by the power utility designated to connect the first turbine, Willenbacher was able to find eight other 'crazy people' to split the one-million Deutschmark investment costs. Willenbacher's 'wind mill' also secured a state grant and received the backing of a commercial bank loan, leaving the nine investors to contribute the residual one-fifth of the funding. The turbine was eventually sited on his uncle's farm, rather than that of his parents, owing to its superior wind resource – but still in the hamlet of Schneebergerhof, southwest of Mainz, where he grew up.

In May 1996, little over nine months after reading the newspaper article from his hospital bed, Willenbacher had his first wind turbine. In the process, he also met Fred Jung, who had similar ambitions for his family farm and with whom he struck a strong rapport. The two subsequently founded juwi – 'ju' for 'Jung' and 'wi' from 'Willenbacher' – which is today a multinational renewable-energy business, specialising in wind and solar project management and engineering, procurement and construction services. As of 2017, juwi has supported the development of 900 wind turbines and 1 600 solar projects worldwide.

Despite resistance from the incumbent utilities, Germany's renewables industry nevertheless gained significant momentum in the subsequent decades. The term 'Energiewende', or energy transition, captured the imagination of many other energy entrepreneurs and began to symbolise not only a favourable change in mindset towards renewables, but also the democratisation of the electricity sector. By 2010, Energiewende even obtained the status of official long-term policy, with firm objectives and targets set for carbon-emission reductions, renewable-energy deployment, energy efficiency and supportive changes to market design.

By 2014, Willenbacher had been so convinced by the progress he had witnessed, he decided to make his 'indecent proposal' to chancellor Angela Merkel, calling on her to support a decentralised-transition model, which he argued could deliver a 100% renewable power system by as early as 2020 – far more ambitious than the state's goal of 35%. 'Please accept my offer', Willenbacher begged Merkel. 'I will donate the shares in my company to Germany's energy cooperatives in exchange for a swift 100% Energiewende'.[2]

An indecent proposal no longer

That this 'indecent proposal' could have feasibly been made at all, however, is the consequence of a far earlier event; one that set in motion the energy transition advocated by people such as Willenbacher. The event in question was the oil crisis of 1973–74, when prices surged and supply came under severe strain after the 12 members of the Organisation of the Petroleum Exporting Countries (Opec) agreed to impose an embargo on oil exports to the US, following president Richard Nixon's decision to support Israel against Egypt in the Yom Kippur War. These supply reductions also affected nearly every European country that relied on Opec imports, despite the fact that only the US, the Netherlands and a few other nations were specifically embargoed.[3] Even though the embargo was lifted only months after its imposition, the economic and policy effects were profound and long lasting. Globally, policymakers began to place a premium on energy independence and governments became more willing to support research into alternative energy technologies.

For France, which in the early 1970s was heavily dependent on energy imports and which had limited coal resources, nuclear was the remedy. In the wake of the oil crisis, France's ambitious nuclear programme resulted in the construction of more than 50 reactors in 20 years. Many other countries turned to their coal resources. In the case of South Africa, which was facing the additional threat of trade sanctions as a result of apartheid, coal was not only the go-to mineral for shoring up electricity independence, but also to bolster liquid-fuel security, using the Fischer-Tropsch coal-to-liquids process.

In countries such as Germany, the US and Japan, part of the research effort was directed towards renewable energy, such as solar thermal, solar photovoltaic (PV) and wind. Technological advances made during this period still form the basis for the current portfolio of renewable-energy technologies. There have, naturally, been efficiency advances in PV panels, while the scale of wind turbines has increased dramatically. However, technology innovation has not been the main cause of the disruption under way in international power systems. Instead, it is the result of the actions taken by certain governments to encourage the integration of wind and solar into their electricity systems. Until recently, such integration relied on subsidies. However, the commercialisation and mass manufacture of solar panels and wind turbines has resulted in a precipitous fall in costs, which is increasingly enabling countries to pursue electricity decarbonisation in the absence of subsidies. Nevertheless, it should also not be forgotten that the initial adoption of renewables was driven by a desire for greater energy independence rather than by climate commitments, or a desire to improve the environmental and/or health outcomes.

In the US, some states set specific targets for the level of renewable energy that should be included in a utility's portfolio. These portfolio standards were associated with specific timelines for increasing the share of renewables. The way the utilities implemented these portfolio standards was, and still is, usually through long-term offtake guarantees with negotiated tariffs for the electricity produced, termed 'Power Purchase Agreements' (PPAs). The lifetime of such PPAs is usually aligned with the economic life of the underlying asset, be it a wind farm or a solar plant. The owner of the asset, called 'Independent Power Producers' (IPPs), finance the cost of the assets on the back of these long-term PPAs. In addition, federal government tax incentives, linked to both investment and production, provide additional support. The Production Tax Credit, whereby renewables generators earned a tax credit for every kilowatt-hour produced for the first ten years of operation, emerged as a major driver. The higher the level of production, the larger the tax credit, which made investing in wind farms popular among equity investors who could offset the credit against profits earned in other sectors.

In Europe, a number of countries adopted feed-in tariffs (FIT), which also offered a guarantee of payments to renewable energy developers for the electricity they produced – similar to the PPAs on the other side of the Atlantic. The two main differences to project-specific PPAs were these: first, the standard-offer-nature of the FIT. The tariff level was not determined on a project level, ideally reflecting the underlying project-specific cost, but rather administratively set by the relevant ministries and the same tariff was applicable to all projects of a specific technology. Second, in a FIT regime, the new-build volume is generally not limited. Whereas in a

PPA regime only the volumes that are negotiated with or auctioned by the off-taker utility will be implemented. A FIT regime, on the other hand, allows any projects that is ready to be connected to the grid.

After a decade or so of small-scale solar PV and wind demonstration projects, Germany adopted the Electricity Feed Act in 1990, followed, in 2000, by a more robust law, known as the Electricity Resources Act (EEG).[4] The EEG refined the feed-in structure and mechanism by, among other things, establishing technology-specific prices, guaranteed for 20 years, for renewable energy. This provided the necessary cost recovery and planning security for investors and greatly boosted the instalment of solar PV plants and wind farms. Through the legislation, grid operators were not only compelled to connect the plants to the network, but also to dispatch the electricity in preference to the energy arising from conventional sources, such as coal, nuclear and gas. No cap was placed on the level of deployment, but the feed-in model was designed so that tariffs for new installations would decrease over time and at regular intervals. The feed-in tariffs, together with a mechanism for reducing the tariffs over time, were set by the policymaker, initially the Ministry for the Environment and, currently, the Ministry for Economic Affairs and Energy. The feed-in tariff model remains in place for the German rooftop PV market, but, since 2016–17, the country has transitioned to an auction model for utility-scale PV and wind farms, while retaining the 20-year offtake guarantee.

To pay for these initially more expensive power stations, electricity consumers have been charged an additional fee, known as the EEG surcharge, named for the Act's German title, 'Erneuerbare-Energien-Gesetz'. To sustain the competitiveness of the energy-intensive industry, companies meeting specific demand thresholds, which are heavily geared towards export markets, can apply for an EEG surcharge exemption. As a result, electricity-intensive firms enjoy tariffs that are lower than those across the border in France,[5] even though other German consumers pay a premium, largely as result of the EEG surcharge and other taxes and levies in the tariff. The reason for this is that renewables have reduced the wholesale electricity price, as determined by the fuel cost of the most expensive operational plant. This is because, as the proportion of wind and solar PV (which have no fuel costs) rises, the 'merit order' shifts to the right and makes supply sources with high fuel costs, such as gas and diesel, uneconomic. For residential and commercial customers, the reduction in wholesale power prices is insufficient to offset the EEG surcharge. But for those heavy industries exempt from the surcharge, their competitiveness has actually improved with the introduction of renewables.

Arguably one of the most powerful aspects of the German feed-in model was the system's built-in bias both towards electricity-industry democratisation and towards smaller installations, such as the wind project initially pursued by Jung and Willenbacher. Under the EEG's feed-in tariff regime

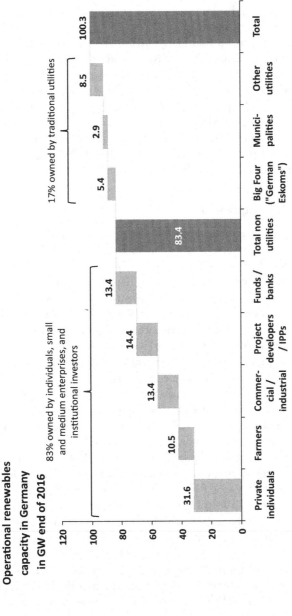

Figure 2.1 Ownership structure of renewables assets in Germany at the end of 2016, excluding hydro. The 'Big Four' utilities are E.ON, RWE, EnBW and Vattenfall

for solar PV, the smaller the project, the higher the tariff received. In addition, the utilities could not undermine deployment on the basis of grid access, as projects were considered to be commercially available on receipt of a certificate stating that the facility was ready to produce, even if it was not yet connected to the network.

These incentives meant that many private individuals, no matter their political allegiance, became direct investors in, and beneficiaries of, the electricity transition. This had the effect of helping to insulate the renewables roll-out from the criticism that inevitably arose when electricity prices began to rise, particularly as a result of a significant acceleration in solar PV installations from 2004 onwards. At the end of 2016, private individuals, farmers, SMMEs[6] and institutional investors owned 83% of the entire non-hydro renewable capacity in Germany of more than 100 GW, underpinning ongoing multiparty support for the continuation of the programme.[7]

Unintended consequences

The 'solar surge' in Germany arose as a result of an unexpected decoupling in the pace at which solar panel prices were falling globally – as panel manufacturing migrated to China – and the far slower pace at which the regulated feed-in tariffs were declining. The gap made for an extremely attractive investment case, which was reflected in 2005 when Germany became the first country in history to install more than 1 GW of solar PV capacity in a single year. Prior to this decoupling, onshore wind and biomass projects (primarily based on maize) had been the main contributors to Germany's renewables expansion – wind for its early relative cost competitiveness and biomass for its high capacity factor.

Another consequence of the two-speed market, where government-determined feed-in tariffs failed to adjust quickly enough to plummeting costs, was market bubbles. In several European markets where feed-in tariffs were designed to decline gradually, investors moved to exploit the opportunity created as PV costs fell precipitously. Spain, Italy and Germany all experienced a surge in investment between 2008 and 2012. In Germany, 22.5 GW of solar PV capacity was added in only three years, while Italy added a whopping 9.3 GW in 2011 alone. The Spanish authorities reacted by stopping not only the feed-in scheme for new installations, but also by retrospectively cutting net tariffs for existing power purchase agreements, a move that massively destroyed investor confidence in that country.

The German regulator took steps to recalibrate its feed-in tariffs, but not in time to prevent a solar investment bulge, which peaked in 2012, when 7.6 GW of solar PV, mostly in the form of rooftop installations, was added. All told, between 2000 and 2017 the country invested over €100 billion into solar PV and added a cumulative 42.4 GW of capacity. Its deployment of wind farms also increased, albeit at steadier pace, and Germany's onshore wind capacity now stands at 50.5 GW, and its offshore wind capacity 5.4 GW. Together with 8.0 GW of biomass/-gas and 5.6 GW of hydro, the total renewables capacity in Germany stood at 112 GW by the end of 2017. These capacity additions have resulted in the share of renewables in electricity consumption increasing from 6.5% in 2000 (mainly hydro) to 36.1% in 2017.[8] Because of that, the carbon intensity of electricity production decreased from 640 g of CO_2 per kWh in 2000 to 500 g of CO_2 per kWh in 2017. This fall in carbon intensity has been achieved despite a decision to gradually, and in parallel, phase out all nuclear plants, next to renewables the only other source of carbon-free power generation.

Much of the solar PV capacity was built, though, in the four years between 2009 and 2012, when the cost of solar PV was still very high. It could be argued, therefore, that Germany has paid the solar PV 'school fees' for the rest of the world – investing when prices were high, but creating the conditions for the sharp decline in costs that followed. The extent of this early-adopter premium can be seen in the fact that the first 38 GWp of solar PV added translated into €9.6 billion in yearly feed-in tariff payments. By contrast, the next 14 GWp, to take Germany to its self-imposed 52 GWp solar PV deployment cap under the feed-in tariff regime, will add only €1 billion to this annual bill.

As a consequence, the focus in Germany is shifting to market design, and away from whether or not to add more wind and solar PV, which are accepted as being the cheapest generation sources. One of the issues up for discussion, for instance, is how best to deploy the country's 5.7 GW-strong biogas fleet, which is currently remunerated for the energy the plants produce. In a system requiring flexibility to fill the gaps left when the wind stops blowing, or when there is no sun, it is not optimal to operate the biogas fleet as baseload, as is the case currently. Baseload operation (i.e. continuous operation with constant power output at maximum rated capacity) in a system with large amounts of variable solar PV and wind is not helpful and congests the grid during sunny and/or windy periods. When the German feed-in tariff regime was drafted in 2000, it was not foreseen that 'baseload operation' could become a problematic configuration. To address the problem, the German biogas regime will have to be redrafted to include incentives for more flexible operation of the biogas plants. However, changing contracts retrospectively will also be problematic.[9]

Nevertheless, the German experience is a cautionary tale for new biogas entrants, such as South Africa. When such countries consider ways to exploit the energy opportunity associated with agricultural and municipal waste, flexibility should be a key consideration and should be factored into procurement programmes from the start. Developers should be incentivised to invest not only in a biogas digester, but also in additional gasholders to store the biogas. In addition, they should be guided to include several gas engines, not only one. Instead of a single 1 MW gas engine running at 6 000 full-load hours yearly, there could be six 1 MW engines running at 1 000 full-load hours each. The same amount of biogas would be burnt yearly, but the outcome would be a highly flexible solution with six times the capacity to fill the gaps created by solar PV and wind. A flexible configuration would also not be cost prohibitive, as a biogas plant's main cost lies not with the gas engines and certainly not with the gasholders, but with the feedstock and feedstock handling, as well as the biogas digester. Should Germany, for instance, adopt such a model, its 5.7 GW biogas fleet, which currently produces 32.5 TWh/a[10] of inflexible baseload power, would be revamped to incorporate a far larger capacity of between 30 GW and 35 GW. Such a fleet would still produce 32.5 TWh/a, but would be able to cover reliably almost 50% of the annual German peak demand at any point in time.

With all these renewables school fees having been paid by the first adopters, the rest of the world, including South Africa, is in a good position to take advantage. And recent statistics indicate that many countries are doing just that. With solar PV costs having fallen more than 80% since 2008 and onshore wind costs by more than half (see Figure 2.2), the initial

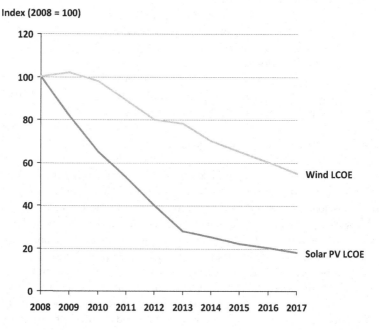

Index (2008 = 100)

Figure 2.2 Global average cost of wind and solar PV power production since 2008, indexed[11]

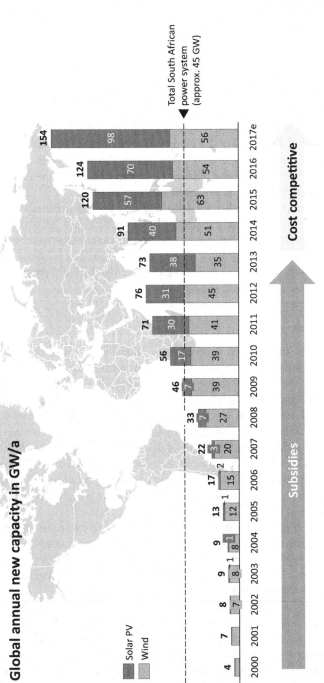

Figure 2.3 Global annual solar PV and wind deployment from 2000 to 2017

Sources: GWEC, Global Wind Statistics 2017, http://gwec.net/wp-content/uploads/vip/GWEC_PRstats2017_EN-003_FINAL.pdf. REN21, Global Status Report Renewables 2018, http://www.ren21.net/gsr-2018/

need to subsidise renewables is rapidly disappearing and the rate of their deployment is accelerating.

During 2017, a record of more than 150 GW of new wind and solar PV capacity was added globally, comprising almost 100 GW of new solar PV plants and more than 50 GW of new wind farms.[12] This represents a material scale-up from the situation that persisted only 17 years earlier when a mere 4 GW of new wind was added in 2000, and solar PV deployment literally did not exist yet (see Figure 2.3). The composition of the new installations has also changed significantly, with wind having played by far the dominant role in earlier deployments, because of its earlier achieved cost competitiveness, and with solar being more prominent in the latter years.

What is also apparent is the fact that there have been two distinct phases to the recent period of rapid expansion. Between 2000 and 2013, deployment rose steadily on the back of policy interventions, supported by subsidies. However, since roughly 2014, the need for subsidies is falling away country by country as the two technologies breached the cost competitiveness tipping point described in Chapter 1.

Pace and scale

This tipping point has been made possible partially by efficiency gains of the solar PV technology. However, the main motive force has been the migration of panel manufacturing to China, which has led to mass manufacturing and fundamentally altered the cost of production. The improvements in wind, meanwhile, are associated with turbine scale, from turbines of around 1–2 MW in the early 2000s to 4-5 MW currently for onshore turbines, longer blades (up to 160 m rotor diameter) and higher hub heights (up to 160 m). A higher hub height gives access to higher wind speeds and hence improves the quality of the harvestable resource, and larger rotor diameters lead to a larger swept area and thus to a higher quantity of harvestable wind resource.

The post-subsidy environment is not only associated with an acceleration in the pace of new builds, however. It is also characterised by considerable geographical diversification. Whereas Europe led the way during the subsidy era, Asia, particularly China, and the US are playing a more prominent role as wind and solar PV emerge as cost leaders.

By the end of 2017, China's installed wind capacity had surged to 188 GW, surpassing Europe's installed wind base of 177 GW by a good margin.[13] China's installed capacity of solar PV, meanwhile, had risen to 131 GW, from almost just zero a few years earlier.

In the age of the 24-hour news cycle, these changes may appear unspectacular, even laborious, particularly when the current contribution of

Table 2.1 Global solar PV and wind operational capacities end of 2017 (estimated)

Country/Region	Wind (end 2017)[1]	Solar PV (end 2017)[2]
China	188 GW	131 GW
Europe	177 GW	113 GW
USA	89 GW	54 GW
Japan	3 GW	53 GW
India	33 GW	19 GW
Americas w/o USA	34 GW	9 GW
Asia Pacific w/o Australia	5 GW	14 GW
Australia	5 GW	7 GW
Middle East and Africa	5 GW	5 GW
. . . *of which South Africa*	*2.1 GW*	*1.8 GW*
Total world	**539 GW**	**405 GW**

Sources: [1] GWEC Global Wind Report 2017, http://files.gwec.net/files/GWR2017.pdf?ref=PR
[2] Solarpower Europe, Global Market Outlook 2017

renewables is assessed within the context of global electricity supply. All non-hydro renewables in 2015 accounted for 7% of worldwide electricity generation, and it is estimated that in 2017 they reached close to 10% (refer to Figure 2.4).

Indeed, Figure 2.4 also shows that since 2000 the only sources of electricity that have grown in terms of relative market share are non-hydro renewables such as solar PV and wind (2% in 2000 to estimated 10% in 2017) and natural gas (18% in 2000 to 23% in 2015). Coal stagnated at 39%, while oil (8% down to 4%) and nuclear (17% down to 11%) declined in their share of global electricity supply.

It should be recognised, though, that the introduction of renewables at scale is a very recent development. Roughly 80% of the globally existing solar PV capacity has been installed only during the last five to six years.[15] In addition, energy assets, by their nature, have long lifespans, which means the transition to new generation technologies takes time. By 2040, Bloomberg New Energy Finance is forecasting that wind and solar will account for 48% of installed capacity and 34% of electricity generation worldwide.[16] That seemingly bullish estimate may even prove conservative, especially considering that solar PV and wind have both individually outpaced nuclear in terms of globally installed capacity already. And that within a record time of five to ten years of deployment. Even the historically unique rapid build-out of the world's nuclear fleet in the 1970s and 1980s did not reach the pace that we have seen in the last decade with respect to the deployment of solar PV and wind (refer to Figure 2.5).

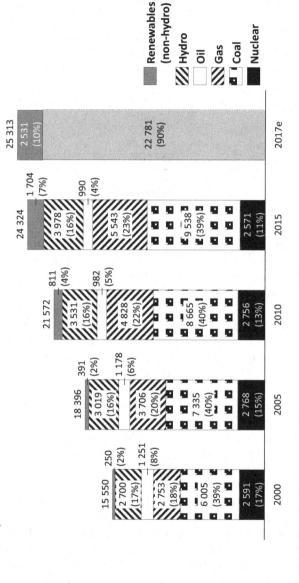

Figure 2.4 Global electricity production from 2000 to 2015 by source, estimated 2017 renewables share[14]

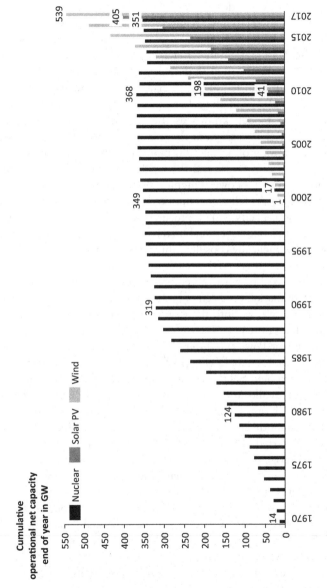

Cumulative operational net capacity end of year in GW

Legend: Nuclear, Solar PV, Wind

Values shown: 14, 124, 319, 349, 368, 539, 405, 351, 198, 41, 17, 1

Years: 1970, 1975, 1980, 1985, 1990, 1995, 2000, 2005, 2010, 2015, 2017

Figure 2.5 Global cumulative, operational nuclear, solar PV and wind capacity[17]

Closer to home

South Africa's renewables deployment also shows signs of both the subsidy and the post-subsidy era, despite the fact that South Africa's renewables journey only began relatively recently. Initially, the country envisaged introducing renewables through technology-specific feed-in tariffs determined by the National Energy Regulator of South Africa. However, policymakers grew concerned that such a model would fall foul of the country's constitutional requirement that all public procurement be the outcome of competitive bidding. Feed-in tariffs, therefore, made way for a reverse-auction model, known as the Renewable Energy Independent Power Producer Procurement Programme (REIPPPP). In fact South Africa was something of an auction pioneer, with Europe, at the time, still relying on feed-in tariffs and the US largely driven by negotiated – rather than auctioned – Power Purchase Agreements (PPAs).

The REIPPPP model relies on a ministerial determination, outlining how much generation capacity will be procured and how that capacity will be allocated across the various renewables technologies. This allocation is itself derived from the publically consulted Integrated Resource Plan for electricity, stating how much generation capacity, renewables and conventional, should be added, and by when, to ensure supply meets forecasted demand. Once the determination has been published, specific technology allocations are set aside and released into the market through a series of competitive bid windows. The entire process is overseen by a government agency known as the Independent Power Producer (IPP) Office, which is a joint venture between the Department of Energy (DoE) and the National Treasury.

The IPP Office held South Africa's inaugural REIPPPP bid window in August 2011 for an allocation of just over 1 400 MW. Bids were submitted in November 2011, and the identities of the first 28 preferred bidders were released just a month later, following an adjudication process involving 53 bids. The preferred bidders named included 18 solar PV projects, eight onshore wind projects and two concentrated solar power projects.[18] All the projects received 20-year PPAs, with the electricity bought by Eskom, which was entitled to recover the cost of such purchases from consumers through the regulated tariff. IPP costs are treated as a pass through in Eskom's revenue applications. The entire programme is backed by a R200 billion National Treasury–approved guarantee, known as the Government Support Framework Agreement. That would be triggered if Eskom was unable to honour the PPA payment obligation and it essentially reduces the off-taker risk from an Eskom to a South African government risk.

The first bid window was hailed as a major success, as it helped kick-start the development of an industry that attracted nearly R200 billion of foreign and domestic investment over the subsequent five years. The REIPPPP model also became a global yardstick, with several countries soon following in South Africa's reverse-auction footsteps. In hindsight, though, the capacity set aside for the 2011 bid window was too large, as it came too early to capture the full benefits of rapidly falling technology costs, especially of solar PV. As a result, the solar PV projects procured in 2011 were awarded 20-year PPAs at a high tariff level of R3.65/kWh (in 2016 prices), while the wind projects came in at R1.51/kWh.

That said, the first successful auction served to raise the profile of South Africa as a credible and relatively low-risk renewables market among investors, which allowed South Africa to begin reaping the benefits of falling wind and solar PV costs. By the second bid window, solar PV tariffs had fallen to R2.18/kWh, while those for onshore wind declined to R1.19/kWh. The decreases continued into bid windows three – to R1.17/kWh for solar PV and R0.87/kWh for wind – and four, when tariffs fell to R0.87/kWh and R0.69/kWh for solar PV and wind respectively.[19]

Just before a backlash from Eskom, which began to publically question the affordability of the REIPPPP and refused to sign new PPAs in 2016, the downward price trend was reaffirmed in November 2015. Owing to the impasse with Eskom, the preferred bidders for the bid window were not immediately identified; however, the DoE revealed that tariffs had declined to an average of only R0.62/kWh for both solar PV and wind. This represented a massive decline in actual tariffs achieved through the REIPPPP; 83% in the case of solar PV and 59% for onshore wind from the first bid window in November 2011 to the most recent in November 2015. A subsequent comparative analysis, based on information provided by the IPP Office, showed that, at those levels, solar PV and onshore wind projects were 40% cheaper than new baseload coal-fired power stations, which cost R1.03/kWh (refer to Figure 2.6).[20]

The stalemate in South Africa between 2015 and 2017, which was only broken in early 2018, was also not reflective of global trends. Both solar PV and wind remained, and still remain, on strong growth trajectories, while their future costs are forecasted to fall even further. Bloomberg New Energy Finance expects renewables will comprise 72%, or $7.4 trillion, of the $10.2 trillion in new power generation investment worldwide to 2040. Of this, solar PV investments are expected to comprise $2.8 trillion and wind $3.3 trillion, with investment in renewables increasing to around $400 billion a year by 2040.[21] In addition, the levelised cost of electricity from new solar PV is forecast to fall by another 66% by 2040, while that from onshore wind will fall by close to 50% over the same period.[22]

These represent significant changes over a relatively short period of time and further strengthen the argument that the world is just at the very beginning of what is shaping up to be a fundamental energy market transition.

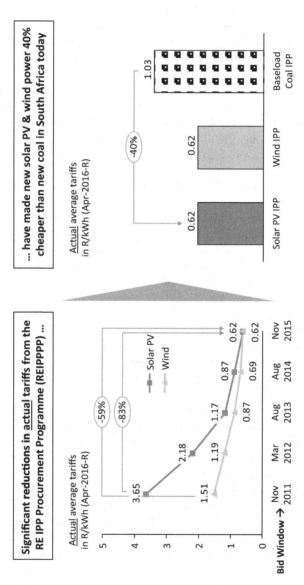

Figure 2.6 Reduction of tariffs in the South African REIPPPP for solar PV and wind, and a comparison with auctioned tariffs for new baseload coal power stations from IPPs

Bottom Line: The tipping point in the cost competitiveness of wind and solar PV is the result of 'school fees' paid by developed countries, which subsidised the deployment of renewables for decades. Follower countries, including South Africa, have an opportunity to invest at close to the bottom of the technology-cost curve and without need for subsidies. Both solar PV and wind have taken over the dominant role in global new power-generation capacity deployment, now purely on the back of their superior economics.

Notes

1   Willenbacher, M. *My Indecent Proposal to the German Chancellor: Because We Cannot Let the Energy Transition Fail!*, 2014.
2   Willenbacher, M. *My Indecent Proposal to the German Chancellor: Because We Cannot Let the Energy Transition Fail!*, 2014.
3   Tonini, A. *The EEC Commission and European Energy Policy: A Historical Appraisal*, 2016.
4   Federal Ministry for Economic Affairs and Energy (BMWi), accessed November 2017, www.erneuerbare-energien.de/EE/Redaktion/DE/Dossier/eeg.html?cms_docId=73930.
5   European Commission, *Quarterly Report on European Electricity Markets*, third quarter 2017. https://ec.europa.eu/energy/sites/ener/files/documents/quarterly_report_on_european_electricity_markets_q3_2017_finalcover.pdf.
6   SMMEs: Small, Medium and Micro-Sized Enterprises.
7   *Trendresearch: Eigentümerstruktur: Erneuerbare Energien.* www.trendresearch.de/studien/20-01174.pdf?65b80e3ed98f1cd7d75ae799433d5b0a.
8   *Agora Energiewende: The Energy Transition in the Power Sector: State of Affairs in 2017.* www.agora-energiewende.de/fileadmin/Projekte/2018/Jahresauswertung_2017/Energiewende_2017_-_State_of_Affairs.pdf.
9   *The Boston Consulting Group (BCG): Towards a Zero-Carbon World – Can Renewables Deliver for Germany?* www.bcg.com/documents/file108854.pdf.
10 German Federal Ministry for Economic Affairs and Energy (BMWi), *Statistics of Renewables in Germany.* www.erneuerbare-energien.de/EE/Navigation/DE/Service/Erneuerbare_Energien_in_Zahlen/Zeitreihen/zeitreihen.html.
11 *IEA: Next Generation Wind and Solar Power, from Cost to Value.* www.iea.org/publications/freepublications/publication/NextGenerationWindandSolarPower.pdf, extrapolated beyond 2015 based on own experience.
12 CSIR analysis using information sourced from the International Energy Agency, *The Global Wind Energy Council, the European Photovoltaic Industry Association, and Bloomberg New Energy Finance.*
13 CSIR analysis using information sourced from the International Energy Agency, *The Global Wind Energy Council, the European Photovoltaic Industry Association, and Bloomberg New Energy Finance.*
14 IEA *Statistics.* www.iea.org/statistics/statisticssearch/report/?country=WORLD&product=electricityandheat&year=2015; *REN21: Renewables 2017 Global Status Report.* www.ren21.net/wp-content/uploads/2017/06/17-8399_GSR_2017_Full_Report_0621_Opt.pdf; *BP Statistical Review.* www.bp.com/content/dam/bp/en/corporate/pdf/energy-economics/statistical-review-2017/bp-statistical-review-of orld-energy-2017-electricity.pdf.

15 CSIR analysis using information sourced from the International Energy Agency, *The Global Wind Energy Council, the European Photovoltaic Industry Association, and Bloomberg New Energy Finance.*

16 Bloomberg New Energy Finance, *New Energy Outlook 2017*, June 2017.

17 World Nuclear Association – Reactor Database; Solarpower Europe; GWEC.

18 Creamer, T. Engineering News, *SA Unveils the Names of First 28 Preferred Renewables Bidders*, December 2011.

19 South African Department of Energy, *Market Overview and Current Levels of Renewable Energy Deployment*, p. 27–28, http://www.energy.gov.za/files/renewable-energy-status-report/Market-Overview-and-Current-Levels-of-Renewable-Energy-Deployment-NERSA.pdf.

20 CSIR Presentation, *Actual Tariffs: New Wind/Solar PV 40% Cheaper than New Coal in RSA*, 2016, http://www.ee.co.za/wp-content/uploads/2016/10/New_Power_Generators_RSA-CSIR-14Oct2016.pdf.

21 Bloomberg New Energy Finance. *New Energy Outlook 2017*, June 2017.

22 Bloomberg New Energy Finance. *New Energy Outlook 2017*, June 2017.

3 It's all about solar and wind

An essential characteristic of any power system is that supply must meet demand everywhere in the system and at every moment. To do so, energy policymakers and system planners typically start by assessing what resources are available to do the job and to do so most cost efficiently.

Until recently, the lowest-cost resource available to South Africa has undoubtedly been coal; an energy mineral that the country has in abundance. Indeed, South Africa's extensive coal resource is not only the basis for its large-scale and robust electricity system (roughly 90% of South Africa's electricity is generated from coal), but even plays a disproportionate role (over 70%) in the larger energy system, owing to the conversion of coal, by Sasol, into liquid fuels using the Fischer-Tropsch technology.

This powerful coal heritage has underpinned the development of South Africa's so-called mineral and energy complex that still helps shape the contours of the country's contemporary economy. It also makes it easy to forget that this was not always the case. Although a small, low-quality coal resource was discovered in 1699 in the Franschhoek Valley, near Stellenbosch, the British were reportedly dismayed at the near absence of the commodity when taking occupation of the Cape in 1814. By 1854, the then colonial administration was even offering a £100 reward for the discovery of a quality coal resource that could be mined at a lower cost than the mineral could be imported.[1] A corresponding editorial in the *Cape Monitor* went on to describe coal as the greatest source of national wealth to a state: 'Gold, silver and copper may enrich individuals . . . but coal diffuses wealth throughout the country, and sheds comfort and prosperity all around', the newspaper article declared. 'The discovery of coal and iron in abundance', the article continued, 'would be of infinitely greater importance, in an industrial point of view, than gold, silver or precious stones'.[2]

With the discovery of diamonds and gold in the latter part of the nineteenth century, demand for fuel, primarily wood, soared as steam power became the main source of motive power for industry around the Kimberley diamond fields. With no workable coal deposits in close proximity, wood was being sourced from as far away as modern-day Botswana and, by the early 1880s, the price of wood was so high that fuel rose to constitute about 30% of total working costs for diamond mine operators.[3] In the event, however, massive quantities of the energy mineral were subsequently discovered. These were mostly located in the north-eastern regions, primarily in what today constitutes the Mpumalanga province. The size of the national resource is currently estimated to be some 30 billion tonnes, or 3.5% of global resources.[4]

South Africa mines about 240 million tonnes of coal yearly, the bulk of which is used domestically to fuel Eskom's fleet of 15 coal-fired power stations, with the new mega Medupi and Kusile coal-fired plants currently in the process of being integrated. However, South Africa is also the sixth-largest exporter of thermal coal, primarily through the Richards Bay Coal Terminal, in KwaZulu-Natal. In addition, Sasol mines some 40 million tonnes of coal a year for gasification and conversion into liquid fuels.

The coal industry is, therefore, not only the bedrock of South Africa's mining, energy and industrial sectors, but also a large employer and foreign-exchange earner. In 2016, coal sales were recorded at around R112-billion, with exports representing close to 50%, or R50 billion of total sales by value. Almost 60% of all exports went to India in 2016.[5] At the same time, the industry employed almost 80 000 people, or about 17% of the total mining-sector workforce. For this reason, much attention will need to be given to managing the social and economic impacts on the coal sector as the energy transition gains momentum, driven not only by the rising competitiveness of wind and solar, but also by South Africa's international climate commitments and by the risk of demand for South African export coal shrinking.

Game changer

The steep fall in the costs associated with solar photovoltaic (PV) and wind turbines has brought two of South Africa's other impressive natural resources – solar and wind – into sharp relief. Previously, exploiting these resources was viewed as an uneconomical pipe dream. However, the fact that electricity from new-build solar PV and onshore wind is now 40% cheaper than that from new-build baseload coal,[6] is a game changer. Indeed, the prospect of transitioning to an electricity system where renewables generation, rather than coal, is the workhorse has become serious.

Few would have any arguments about the potency of South Africa's solar resource. 'Sunny South Africa', as citizens and visitors alike often describe the country, is recognised as having almost unparalleled solar irradiation, which is still largely untapped (South Africa has almost two times the solar resource of Germany, where around 40 GW of solar PV is already installed).

A joint analysis conducted in 2016 by the CSIR, Eskom, the South African National Energy Development Institute, Fraunhofer IWES (now IEE) and Technical University of Denmark (DTU), of Germany and Denmark respectively, offers some insight into just how large South Africa's solar resource is. Using only existing environmental impact assessment applications for utility-scale solar PV projects, the study found the combined capacity associated with the applications to be equivalent to some 220 GW – an installed base that could yield about 420 terawatt-hours (TWh) a year of electricity. In addition, using conservative estimates for rooftop solar penetration, the researchers argued that a further 73 GW of distributed solar PV could be added – the equivalent of 136 TWh a year of electricity. Together, therefore, rooftop and utility-scale solar PV showed a conservative total potential of nearly 300 GW of installed capacity or more than 550 TWh of yearly electricity generation.[7] In addition, solar PV supply in South Africa has low seasonal variations, which is not the case in a number of countries where the potential to produce electricity from solar irradiation falls sharply during the winter months. To put the 550-TWh/a figure into perspective, South Africa's total electricity demand stands at around 250 TWh/a, including domestic customer demand, exports and grid losses.

Scepticism remained, however, about the quality and distribution of South Africa's wind resources. While there is general acceptance that the coastal areas of the Eastern Cape and Western Cape provinces have good potential, few would immediately describe the resources in the rest of the country as being world class. However, the same 2016 analysis does just that. Using Wind Atlas South Africa data, the researchers simulated wind power across South Africa in a very high 5 km by 5 km spatial resolution, or almost 50 000 pixels. Using five years of temporal coverage from 2009 to 2013 at a 15-minute temporal resolution, the study found that, at 100 m above ground (the height at which most turbines harvest wind energy), wide areas of South Africa have average winds speeds above 6 m/s – the minimum average wind speed that is required by wind farm designers and operators to make a feasible business case.[8]

For a wind turbine's ability to harvest the wind, the two most important parameters are the hub height and the blade diameters. Generally speaking, the higher the hub height and the larger the turbine blades, the more energy can be extracted from the wind at a certain site, all other things being equal. Higher turbines with longer blades are always more expensive than lower turbines with short blades. But if these turbines yield 40% capacity factors as compared to 20%, the higher investment cost might make sense. The optimal turbine parameter choice is, therefore, a site-specific economic-optimisation exercise. The researchers therefore overlaid the wind speed findings with five different generic wind-turbine types to assess what capacity factors could be achieved. The turbines had hub heights ranging from 80 m to 140 m, blade diameters of between 90 m and 117 m and nominal power ratings of between 2.2 MW and 3 MW.

The outcome was impressive, with the simulation indicating that, for the period of 2009 to 2013, capacity factors of 30% were achievable almost everywhere across the country, with particularly strong results for the Western, Eastern and Northern Cape provinces, but also with feasible wind sites in provinces that were previously not on the radar: North West, the Free State, Mpumalanga, Limpopo and especially Kwazulu-Natal. The research team describes the performance as world class, noting that it was superior to capacity factors of 25% and 20% being achieved in Spain and Germany respectively. In fact, they were comparable to the North American fleet, which operates mainly in the windy belt from the Canadian Prairies to northern Mexico. The 2016 study further concluded that, on almost 70% of suitable land area in South Africa, a 35% capacity factor or higher could be achieved. Nevertheless, a well-planned spatial distribution would help materially in smoothing the generation profile. For instance, while a single wind farm could have rapid changes to its power output, this profile improved considerably when aggregated across multiple sites. Aggregating across just ten wind farms dramatically reduced short-term fluctuations, while aggregating across 100 sites reduces 15-minute gradients to almost zero.[9]

Short-term gradients in the supply profile of power generators are a key concern for system operators, especially if they are difficult to forecast. If a power generator reduces its output by a large amount in a short space of time, other power generators need to ramp up and/or flexible customer loads need to be reduced in order to keep the instantaneous supply and demand balance intact. Historically, large power systems always had to cater for intermittent or discontinuous events, such as the collapse of a coal silo that supplies a coal-fired power station with fuel, or the collapse of a large transmission line. Such events cause a step-function-wise supply profile. A spatially disbursed fleet of wind turbines (and solar PV plants, for that matter) never produce such a step function. That is why wind and solar PV are called variable power generators, rather than intermittent sources of supply.

In a 1999 paper, the National Renewable Energy Laboratory (NREL) in the US analysed the correlation of 5-minute gradients of the wind supply profile for spatially disbursed actual wind turbines.[10] The result was that the 5-minute gradients are not correlated as soon as the distance between two wind turbines is more than 20 km. Hence, it is impossible for a fleet of wind turbines to have very steep 5-minute gradients. The reason for this is meteorology and how wind moves over space. It takes time for a wind-speed front to move, and short-term gradients can, therefore, physically not be correlated if the locations are far apart. As a result, short-term gradients are not a concern if the interconnected wind fleet is spatially distributed over wide geographical areas.

In fact, the 2016 South African study found solar PV, rather than wind, to be the main driver for 15-minute gradients in the residual load, owing to the astronomical movement of the sun, which results in a bell-shaped output curve. However, these solar-related gradients are highly predictable and therefore not a major source of concern for system operators either.

The quality of South Africa's solar and wind resources has been verified by the performance of the projects that have been brought into commercial operation as a result of the government's Renewable Energy Independent Power Producer Procurement Programme (REIPPPP). Since the introduction of the plants in late 2013, actual solar PV plants have been operating with 25% capacity factor (compared to 10% in most parts of Central Europe) and onshore wind projects have been operating with average capacity factors of more than 35% (compared to 20–25% in Europe for comparable hub heights and blade diameters).[11]

Powerful combination

These positive characteristics increase even further when South Africa's wind and solar resources are combined into a composite portfolio, with the model showing that high overall shares of renewable energy can be realised with low short-term variability. In a generation mix where the share of variable renewable energy (VRE), such as solar PV and wind, is 30%, no significant increase in 15-minute gradients of the residual load can be detected; wind alone could provide 50% of the total energy demand without a significant increase in the 15-minute gradients. In addition, the 2016 study argues that a 65% VRE share can be achieved with almost no excess energy and that electricity storage to absorb excess electricity is only required at a VRE share above this threshold. This excess would be derived mainly from solar PV, which, beyond a certain threshold, causes large amounts of excess electricity, owing to the technology's daytime-only supply pattern. Wind supply is more volatile, but on average has a better distribution over the full 24 hours of the day.

For example, to supply a yearly system load of 500 TWh, using a 65% VRE share – comprising 80 TWh of solar PV and 250 TWh of onshore wind – the excess electricity is calculated at only 1.2% of total solar PV and wind energy produced. Such an outcome is low by international standards and is a result of the fact that South Africa is a large country, making the spatial-smoothing effect particularly strong. Explained differently, a wind fleet spread across the country seldom produces at a very high combined output (which is good) and seldom does it produce at a very low combined output either (Figure 3.1).

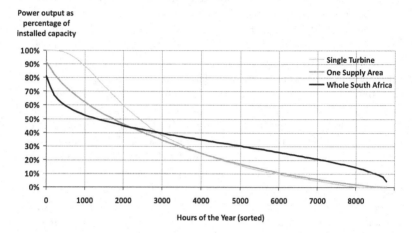

Figure 3.1 Duration curves for a single wind turbine, for a fleet of wind turbines distributed over an Eskom supply area (Johannesburg) and for a fleet of wind turbines distributed over the whole of South Africa[12]

The overall conclusion of the aggregation study was that South Africa exhibits extremely good conditions for both wind and solar energy.[13] Almost the entire country has sufficient resources for profitable wind projects and high capacity factors can be achieved almost everywhere for both solar PV and wind. Spatial aggregation brings major benefits and leverages South Africa's vast land mass. It also shows an average wind output profile that is lower during the day than it is at night, which complements the solar PV profile.

Nevertheless, sceptics often question whether South Africa has enough land to accommodate very high VRE penetration levels. Here again, the 2016 study is definitive in its affirmation that land is not a constraint to a renewables-led generation mix. It concludes that, using only a fraction of its land area, South Africa would not only be able to generate more electricity from wind and solar today than it currently needs, but could also meet all future needs. What's more, the country could do so without displacing alternative productive or social uses for the land. In order to achieve 50 TWh a year, 15 GW of wind-farm capacity would have to be installed, requiring 1 500 km² of land, which is the equivalent of 0.12% of South Africa's land mass. Even scaled up to 250 TWh a year, or a capacity of 75 GW, only 7 500 km², or 0.6% of the country's land mass would be required. Moreover, only a small portion of a wind farm's area is actually utilised land, which means that of the 7 500 km² required for a 75 GW fleet, just 150 km² of that will actually be used for erecting turbines.[14]

If we define the energy-demand density measure by dividing power system size (annual system load) by land area for a number of different countries that already have relatively large wind and/or solar PV deployment, it becomes intuitively clear why South Africa is so perfectly positioned for deployment of energy sources that depend on available land area, such as solar PV and wind. Germany has an energy-demand density of ca. 1 700 MWh/km²/a (600 TWh/a demand divided by 357 000 km² of land area), Italy of 1 000 MWh/km²/a, Spain, Texas and California of 500–700 MWh/km²/a, while the energy-demand density of Ireland is 300 MWh/km²/a. By contrast, South Africa has an energy-demand density of only 200 MWh/km²/a, due to its low population density. This relatively low density on the demand side, combined with higher available resources on the supply side from wind and solar makes it easier for South Africa to implement a high solar PV and wind power system compared to its peers.[15]

Thought experiment: greenfield power system planning

Once it is accepted that South Africa has the solar, wind and land resources to technically host a generation mix led by VRE, attention then, inevitably, shifts to filling the supply gaps that will arise as a consequence of the variable supply profile of the weather-dispatched plants. Equally, the question has to be posed as to whether it is economically feasible to 'back-up' these renewables workhorses with flexible generation technologies that are able to fill the gaps when there is no wind and no sun. In other words, can a solar PV, wind and flexible-power blend meet demand at any given point in time in the same reliable manner as a conventional mix of power generators? If so, would such a mix be cheaper than the alternatives?

Let us conduct a thought experiment. In reality the demand profile in the power system is never purely 'baseload', it varies intra-day, intraweek and seasonally. Nevertheless, let us assume a theoretical, worst-case (from wind/solar perspective), pure baseload demand of constant 8 GW (this translates into 70 TWh a year, which is about 30% of South Africa's prevailing demand) to stress-test the techno-economics of a supply mix led by solar PV and wind.

The constant, 24/7 baseload demand of 8 GW needs to be supplied at any given point in time, reliably. Historically, that would have been done with a fleet of coal-fired or nuclear power stations (6–15 individual units, depending on the size per unit), totalling more than 8 GW installed capacity in order to cater for planned and unplanned downtimes. However, technically the baseload could also be supplied by a mix comprising 7 GW of solar PV and 19 GW of onshore wind, backed up by 8 GW of flexible power generators. The flexible power generator could come in the form of natural gas, biogas, pumped hydro, hydropower, concentrated solar power or demand response.

Figure 3.2 and Figure 3.3 show how this baseload demand can be supplied from a mix of variable supply sources (solar PV and wind), blended with a flexible power generator to fill the gaps. The solar PV and wind supply profiles are the results of a simulation of a 19 GW wind fleet and a 7 GW solar PV fleet spread across South Africa. The simulation was done for the three years from 2010 to 2012. In Figure 3.2, the combined least sunny/windy week is shown, while in Figure 3.3 the combined sunniest/windiest week in this three-year time window is shown.

There are always two distinct times in such a power system: times of undersupply from solar PV and wind power (residual load is greater than zero, no excess solar PV/wind energy), and times of oversupply of solar PV and wind power (no residual load, but excess solar PV/wind energy is greater zero).

During times of undersupply the flexible power generator needs to supply the residual load to make up the total 8 GW of supply that is needed at any given point in time to match the constant 8 GW demand. As long as the flexible power generator is technically able to supply the residual load profile, the customer demand can be supplied at any given point in time, reliably.

During times of oversupply, the excess solar PV and/or wind energy needs to be curtailed. Technically, this is not a problem. The power electronics of solar PV and wind power plants can be configured to produce power at any set point that lies between zero and the maximum available wind/solar resource at that point in time. The curtailment is a pure economic consideration, as curtailed energy needs to be paid for regardless of its use. The wind farms have been built, the solar PV plants have been built, and to not use them does not save any money, because they have zero marginal cost of production.

The outcome shows that it is indeed technically possible for such a mix to supply baseload in as reliable a manner as would be the case using conventional coal or nuclear generators. On an annual average, 89% of the demand of 70 TWh/a would be supplied from solar PV and wind, with the flexible power generators making up the 11% residual.[16] A total of 13 TWh/a, or 17%, of the wind and solar PV energy production of 75 TWh/a would have to be curtailed. Essentially this increases the effective cost of solar PV and wind by $1 / (1-17\%) - 1 = 20\%$, assuming that the curtailed energy cannot be used to supply any additional load on top of the baseload of 8 GW, but has to be 'thrown away'.

Technically, there is no reason why such a power system could not be designed with today's available technologies. The technical feasibility is not a concern anymore. What is much more of relevance is the question of economic viability. From an economic perspective, the following assumptions are made: wind and solar PV both cost 62 c/kWh (South African Randcents per kilowatt-hour); while the flexible power generator is assumed to cost 210 c/kWh. The wind and solar PV costs are derived from the 'Expedited Bid Window',[17] which, at the time of writing, was the most

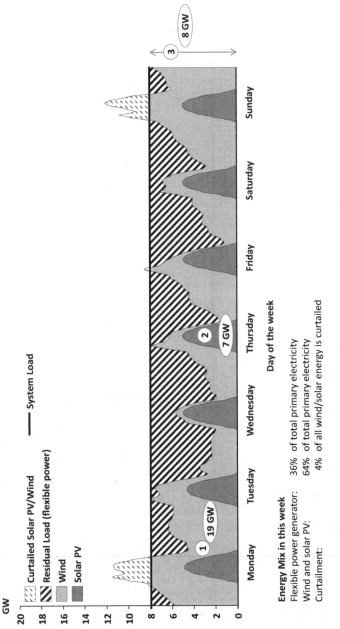

Figure 3.2 Supply of hypothetical baseload demand from 19 GW of variable wind (1), 7 GW of variable solar PV (2) and 8 GW of a flexible power generator (3) for back-up during the least sunny/windy week in the time period from 2010 to 2012 in South Africa

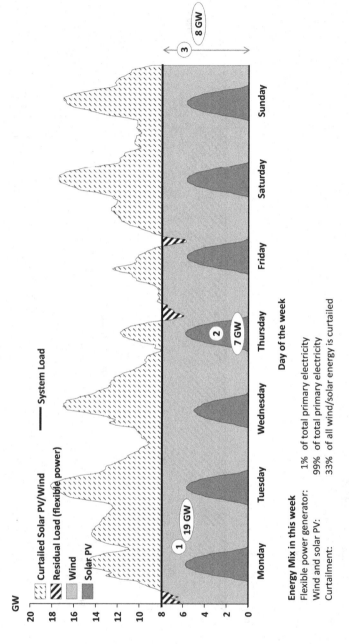

Figure 3.3 Supply of hypothetical baseload demand from 19 GW of variable wind, 7 GW of variable solar PV and 8 GW of a flexible power generator for back-up during the most sunny/windy week in the time period from 2010 to 2012 in South Africa

recent bid window under South Africa's REIPPPP. The cost of the flexible generator is a conservative figure for a power generator with a low utilisation: in this instance, an 11% annual capacity factor. Low-efficiency gas turbines, operated with expensive imported natural gas in form of liquefied natural gas (LNG), at low utilisation, would come in at such a price. As can be seen from the calculations in Figure 3.4, the mix of solar PV, wind and expensive flexible power would cost 90 c/kWh. This number factors in the high cost of supplying electricity during times of no/low wind and sun, and it also factors in the economic cost of 'wasting' 13 TWh/a of solar PV and wind energy that is curtailed. Still, the 90 c/kWh is a good 13% to 17% cheaper than the alternative of supplying the 8 GW baseload demand in the form of either new coal (103 c/kWh) or new nuclear (109 c/kWh). Hence the mix also passes the economic viability test. In fact, under very pessimistic assumptions, the solar PV-wind-flexibility mix is still far superior when compared with coal or nuclear new-build options.

From an environmental perspective, the mix of solar PV, wind and assumed gas-fired power stations emits 64 g/kWh of CO_2, far less than a coal-fired power station that emits approximately 900 g/kWh.

Of course, the demand in this thought experiment can be scaled up from the assumed 8 GW to any other synthetic baseload demand. It does not change the outcome of the economic superiority of the solar-wind-flexibility mix – just more of it gets built. The concept can also be applied to the actual demand profile of South Africa's electricity system, today or in the future. Currently that load profile rises strongly every morning, remains relatively stable throughout the day, before peaking in the evening. The mix of solar-wind-flexibility still costs 90 c/kWh for such a system, while the alternative of building new coal and new nuclear would be substantially higher, because a variable-load scenario is much less favourable for those two technologies than is a synthetic pure baseload scenario. In other words, the real demand profile applied will lead to an even higher cost differential.

An analysis of this thought experiment will no doubt raise a question as to whether it makes sense to supply 8 GW baseload with an installed capacity of 34 GW, comprising 19 GW of onshore wind, 7 GW of solar PV and 8 GW of flexible power generators. The simple answer is yes, because it is about energy, not capacity. True, the rectangular-shaped supply profile of a nuclear or coal plant has a higher value to a power system than a variable supply profile. But by how much? Recent research into that by the CSIR has shown that the value of a rectangular-shaped supply profile is only 10% higher than that of a variable supply profile, up to penetration levels of 40% wind energy share.[18] However, the differential between the LCOE of new solar PV or wind and that of new-build coal is 40%, while the differential to nuclear is almost 50%. Therefore, it is now cost-optimal to meet the electricity demand through variable wind and solar PV, combined with

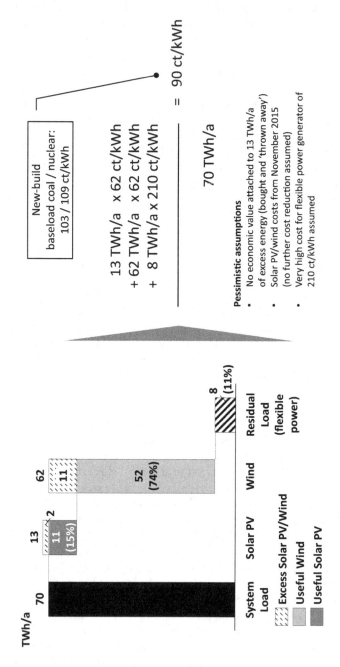

Figure 3.4 Annual energy balance of a hypothetical power system with 8 GW constant demand ('baseload'), i.e. 70 TWh/a energy demand, which is supplied by a spatially disbursed fleet of solar PV, wind and by a flexible power generator to make up the gaps of no/low wind and sun, and resulting average cost of such a supply mix per energy unit

more expensive flexible plants – a reality that will be further reinforced as the cost-competitiveness of these technologies continues to improve.

Real world integrated resource plan: brownfield power-system planning

In the real world of course there is an existing fleet of power generators, which needs to be taken into account when new-build power generators are planned. In South Africa the existing fleet of power generators is dominated by Eskom's coal fleet; Africa's only nuclear power station, Koeberg; some hydropower (most of it imported from Mozambique) and, more recently, solar PV, wind and concentrated solar power (CSP). All the existing power generators, including those to which South Africa has committed, or which are under construction, have a certain economic lifetime. Once they become too old to operate economically they are scheduled for decommissioning. At the same time, the electricity demand is forecasted to increase in South Africa moderately to an estimated 380 TWh/a by 2050 from today's 240 TWh/a. This is a 1.4% growth per annum, which is modest for an emerging economy such as South Africa, where population growth is in the same order of magnitude. Figure 3.5 shows the expected annual energy contribution from the aging existing South African fleet of

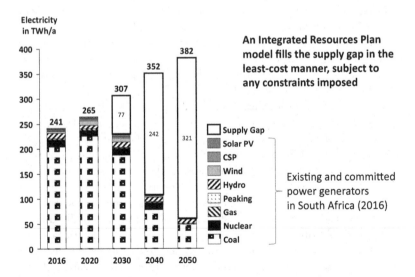

Figure 3.5 Energy contribution from South Africa's existing fleet of power generators from 2016 to 2050, increasing customer demand and widening supply gap from mid-2020s to 2050

power generators, and how the supply gap between demand and existing fleet widens from 2020 to 2050. This supply gap is what needs to be closed by building new power generators. South Africa's Integrated Resource Plan (IRP) seeks to fill this gap in the least-cost manner.

The supply gap widens for two reasons. Firstly, because the demand is expected to grow, with all the uncertainty in such predictions. And secondly, because the existing fleet of predominantly coal-fired power stations will be decommissioned, with large quantities of coal-fired capacity going offline between 2020 and 2040. The decision to decommission plants will be driven by the fact that Eskom's coal fleet is old, environmentally not compliant, and it will simply not be economical to keep them operational. A good part of the increasing supply gap – the one resulting from reductions on the supply-side – is therefore highly predictable.

In order to know how to fill the supply gap in the most economical manner, the technical characteristics and the costs of all available supply-side options need to be collected. Owing to differences in resources (solar, wind, availability of fossil fuels), these costs will differ from country to country. For South Africa, the cost of different new-build supply-side options are very well known for solar PV, wind and new coal-fired power stations in baseload operations, because transparent, competitive auctions uncovered those costs in a highly comparable manner. For all other technologies the Department of Energy made assumptions through market intelligence on technology and fuel-cost side in its Draft Integrated Resource Plan 2016. The different cost of new-build supply-side options are summarised in Table 3.1.[20]

In addition to today's cost, assumptions need to be made about what the future costs will be, especially for those new technologies that are not at the end of their respective maturity and associated cost reduction potential yet. Those technologies are highlighted in Table 3.2. All other costs are assumed to remain as they were in 2016, which is justified because the conventional power-generation technologies are largely matured already over decades of deployment and have little room for additional cost reductions.

What is also highlighted in this table are the two demand-side options for intra-day demand shifting that are most likely to be implemented in the foreseeable future, because it does not require any technological advancements to do so. Firstly, there is the intra-day shifting of the electrical energy demand from the provision of residential warm water, and, secondly, there is the intra-day shifting of the charging of an assumed fleet of electric vehicles in South Africa. Such intra-day shifting of demand would be implemented in a manner that does not affect end-customer utility. That means end customers would not be able to recognise that their devices (warm water boilers and electric vehicles) are dispatched in a system-friendly manner. The power system benefits from the swarm effect of millions of such devices, while the individual device contributes only little to the overall effect.

Table 3.1 Cost assumptions for new-build options in the South African power system in 2016 prices. Solar PV, wind and baseload coal are actuals (results of competitive auctions), while all others are assumptions as per the Department of Energy's Draft IRP 2016

Unit	Operation Regime	Assumed Capacity Factor	Fixed Cost (Capital, O&M) at assumed capacity factor	Variable Cost (Fuel, water, limestone)	LCOE at assumed capacity factor	CO_2 emissions factor
./.	./.	./.	R/kWh	R/kWh	R/kWh	g/kWh
Solar PV	Variable	23%	0.62	0.00	0.62	0
Wind	Variable	40%	0.62	0.00	0.62	0
Coal	Inflexible (baseload)	82%	0.68	0.35	1.03	900
Nuclear	Inflexible (baseload)	90%	0.97	0.12	1.09	0
Gas (CCGT)	Flexible (mid-merit)	50%	0.28	0.87	1.15	400
Coal	Flexible (mid-merit)	50%	1.12	0.35	1.47	900
Gas (OCGT)	Back-up (peaking)	10%	1.09	1.17	2.26	600
Diesel (OCGT)	Back-up (peaking)	10%	1.09	1.98	3.07	600

Table 3.2 Forecasted developments for new energy technologies in constant 2016 prices[19]

Dimension	Technology	Parameter	2016	2030	2050
Primary electricity	Solar PV	Cost per energy unit produced	0.62 R/kWh	0.37 R/kWh	0.20 R/kWh
	Wind	Cost per energy unit produced	0.62 R/kWh	0.46 R/kWh	0.35 R/kWh
Short-term storage	Battery storage	Cost per storage capacity	500 $/kWh	200 $/kWh	100 $/kWh
Short-term demand shifting	Residential warm water	Households with electric warm water	5–6 million (resistive heating)	11 million (partly heat pumps)	27 million (all heat pumps)
		Controllable capacity potential	0 GW	-600 MW +5 GW	-3 GW +24 GW
		Intra-day shiftable energy	0 GWh/d	15 GWh/d	72 GWh/d
	Electric vehicles	Number of electric vehicles	0	1 million	10 million
		Controllable capacity potential	0 GW	-400 MW +4.6 GW	-4.2 GW +96 GW
		Intra-day shiftable energy	0 GWh/d	10 GWh/d	100 GWh/d

Given these assumptions for technology costs and demand-side flexibility options, the supply gap until 2050 is overwhelmingly filled with solar PV and wind power generators. All new power generators in South Africa will be either solar PV, wind or flexible power stations, and built whenever demand increases, or when required as a result of the decommissioning of existing coal-fired power stations. Neither new coal nor new nuclear power stations are part of the least-cost mix. This is because they are more expensive, per kilowatt-hour, than the new bulk energy providers (solar PV and wind) and because they are not flexible enough, both technically and economically, to accompany variable solar PV and wind optimally.

New system workhorses: solar PV and wind

The CSIR, in its latest power-system optimisation, found that it is least-cost to supply 82% of all primary electricity in South Africa from solar PV and wind by 2050. This is shown in Figure 3.6. In this cost-optimal expansion path, roughly 82 GW of solar PV power plants are operational by 2050 and 75 GW of wind farms. They will have the potential to produce a total of 375 TWh of electricity per year, of which 57 TWh are curtailed and 318 TWh are supplying the system load. The fleet of renewable power generators will be accompanied by flexibility options in the form of the residual coal fleet (8 GW, producing 49 TWh/a by 2050), 27 GW of peaking power plants, 5 GW pumped storage and 16 GW stationary batteries. In addition, electric warm-water boilers, or geysers, and electric vehicles will provide flexible capacity through intra-day demand shifting.

No new coal and no new nuclear capacity would be built in the least-cost expansion path. In fact under this path, the last two units of the mega coal project, Kusile, under construction in the Mpumalanga province, would not be finished, because it is more economical to halt construction than to complete them. The 8 GW residual coal plants in 2050 would be all six units of the second megaproject, Medupi (4.3 GW), the four completed units of Kusile (2.9 GW) and one residual unit of Majuba (0.7 GW). By 2050, South Africa's only nuclear power station Koeberg will be decommissioned, with its planned decommissioning date being in the early 2040s.

It is worth mentioning that this significant roll-out of renewable capacities does not require any subsidies whatsoever. It is the least-cost expansion pathway for the South African power system and, therefore, by definition free of any subsidies. Subsidisation would only occur if a policy-driven deviation from this techno-economical least-cost path is pursued. Both in terms of climate-harming carbon dioxide (CO_2) emissions and in terms of water usage, this least-cost pathway will reduce the environmental impact of South Africa's power system significantly. CO_2 emissions, which mostly stem from

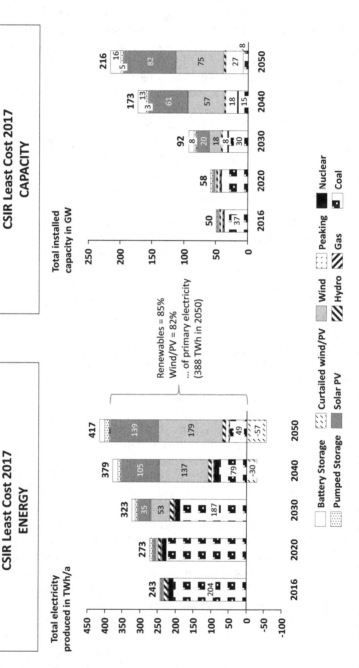

Figure 3.6 Energy and capacity structure of the South African power system in a least-cost expansion plan

coal-fired power stations, will reduce by 74% compared to today, while the fresh-water usage of the electricity sector will reduce by 97%. The reason is that all new power stations are either free of any water use (wind and solar PV) or use very small amounts of water (gas-fired peaking plants) compared with the existing and gradually ramping-down fleet of coal-fired power stations. This is particularly important for South Africa with its increasingly tight water-supply situation.

The curtailed energy from wind and solar PV is fully economically accounted for, i.e. the energy is part of the overall cost of that power system in 2050. If the curtailed energy can be brought to economic use instead of being curtailed, the overall economics of the least-cost expansion path improve even further. Because both wind and solar PV have no fuel and therefore no marginal cost, this curtailed energy is available 'for free' if a suitable demand can be created to absorb it. Such demand could, for example, arise in the form of seawater desalination or the electricity-based production of hydrogen for use in chemical and other industrial processes.

The system optimisation model currently does not allow for the permanent decommissioning of existing coal-fired power stations before their planned shutdown date. The five oldest coal-fired power stations could however be decommissioned even earlier at no or even negative cost.[21] This would make the transition even faster.

In summary, South Africa can decarbonise its power system by 74% by 2050 at no cost. The increasing cost advantage for renewables over coal plays out over time and means that the business-as-usual pathway based on coal is now more expensive than the renewables pathway. In South Africa there is, thus, no longer any trade-off between 'least cost' and 'clean'. The least-cost pathway is at the same time the cheapest and a major step closer to a decarbonised power system.

Figures 3.8 to 3.11 show typical weekly supply profiles in the South African power system in the least-cost expansion pathway. As old coal stations retire, the supply structure is increasingly dominated by variable solar PV and wind, especially from 2030 onwards. It is worth noting that the amount of energy needed from flexible power generators, which are more expensive per kilowatt-hour than the bulk energy providers (today: coal, future: solar PV and wind) and are needed to balance the system, is higher in today's power system at roughly 5%, compared to 3% in 2040 and 2050. The expensiveness of these flexible kilowatt-hours has therefore no effect on the overall cost of the power system, because only very small amounts of them are needed. If the flexible power generators are gas-fired, and that gas is imported, the effect on the trade balance is also negligible for the same reason: the magnitude of the energy needed is simply too small to have any significant effect on the overall cost structure.

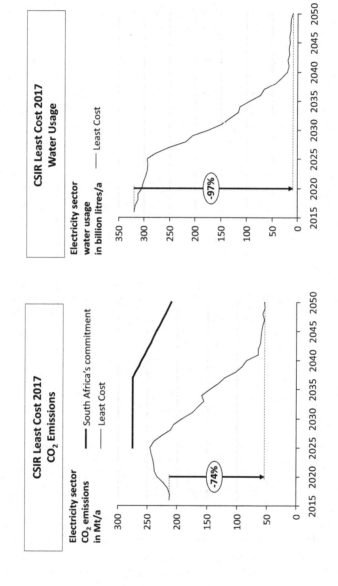

Figure 3.7 CO$_2$ emissions and water usage of the South African power system in a least-cost expansion plan

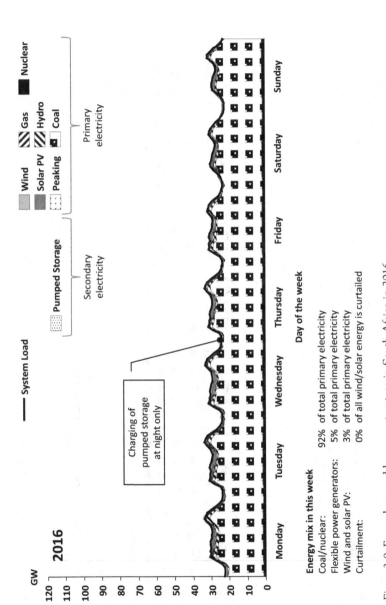

Legend:
- System Load
- Wind
- Gas
- Nuclear
- Solar PV
- Hydro
- Pumped Storage
- Peaking
- Coal

Secondary electricity | Primary electricity

GW

120
110
100
90
80
70
60
50
40
30
20
10
0

2016

Charging of pumped storage at night only

Monday | Tuesday | Wednesday | Thursday | Friday | Saturday | Sunday

Day of the week

Energy mix in this week

Coal/nuclear:	92% of total primary electricity
Flexible power generators:	5% of total primary electricity
Wind and solar PV:	3% of total primary electricity
Curtailment:	0% of all wind/solar energy is curtailed

Figure 3.8 Exemplary weekly power structure in South Africa in 2016

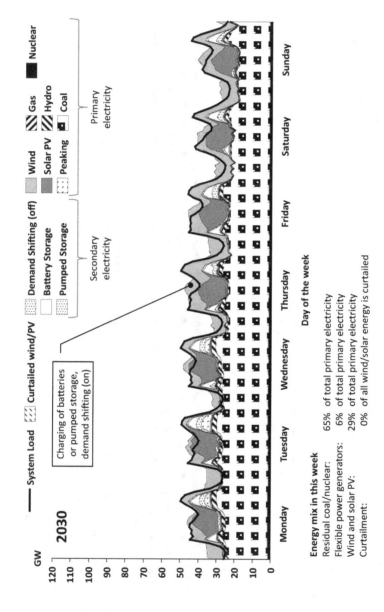

GW

2030

120
110
100
90
80
70
60
50
40
30
20
10
0

System Load   Curtailed wind/PV   Demand Shifting (off)   Wind   Gas   Nuclear
Battery Storage   Solar PV   Hydro
Pumped Storage   Peaking   Coal

Secondary electricity

Primary electricity

Charging of batteries or pumped storage, demand shifting (on)

Monday | Tuesday | Wednesday | Thursday | Friday | Saturday | Sunday

Day of the week

Energy mix in this week

Residual coal/nuclear: 65%   of total primary electricity
Flexible power generators: 6%   of total primary electricity
Wind and solar PV: 29%   of total primary electricity
Curtailment: 0%   of all wind/solar energy is curtailed

Figure 3.9 Exemplary weekly power structure in South Africa in 2030 with least-cost expansion

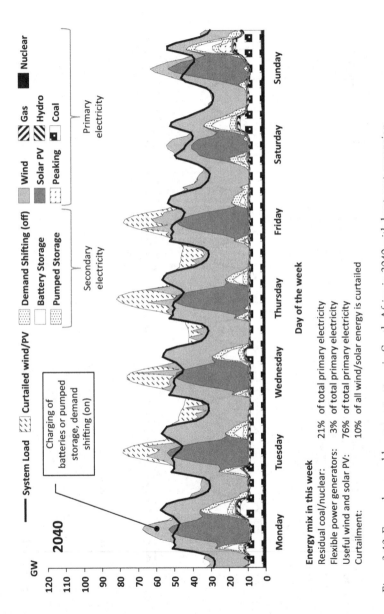

Legend:

System Load — Curtailed wind/PV — Demand Shifting (off) — Wind — Gas — Nuclear — Solar PV — Hydro — Peaking — Coal — Battery Storage — Pumped Storage

Charging of batteries or pumped storage, demand shifting (on)

2040

Secondary electricity

Primary electricity

GW

Day of the week

Monday — Tuesday — Wednesday — Thursday — Friday — Saturday — Sunday

Energy mix in this week

Residual coal/nuclear: 21% of total primary electricity
Flexible power generators: 3% of total primary electricity
Useful wind and solar PV: 76% of total primary electricity
Curtailment: 10% of all wind/solar energy is curtailed

Figure 3.10 Exemplary weekly power structure in South Africa in 2040 with least-cost expansion

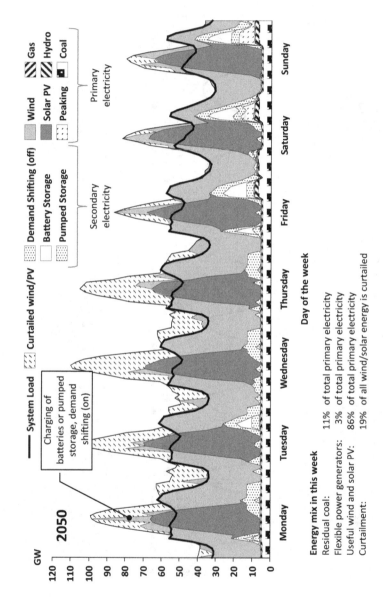

Figure 3.11 Exemplary weekly power structure in South Africa in 2050 with least-cost expansion

It is obviously good news for South Africa that a big step towards decarbonisation of its electricity sector can be achieved without financial effort, simply following the least-cost pathway. Nevertheless, it is worth noting that this least-cost pathway is still not sufficient to achieve the global climate targets as agreed at the United Nation's Conference of the Parties (COP21) in Paris in 2015. There countries made binding commitments to limit global warming to 2 degrees Celsius compared to the pre-industrial age, and to make all efforts to limit it to 1.5 degrees Celsius. In order to achieve this, the International Energy Agency (IEA) calculates that South Africa would have to decarbonise its electricity sector more deeply and faster than would be achieved under the least-cost pathway (see Figure 3.12).[22]

Such a deep-decarbonisation required from South Africa in the IEA's Beyond 2 Degrees Scenario (B2DS) would cost more than least-cost expansion of the power sector, but only if externalities of coal are not internalised in the models – which they are not today. And it may still be that the deep-decarbonised pathway is cheaper than coal-based business-as-usual.

Flexibility critical

Crucially, none of the generic power generators discussed so far is able to supply a typical customer demand in a power system on its own. This is because customer demand is neither rectangular-shaped (i.e. constant all the time), nor does it assume the daytime bell-shape of solar PV or the constantly changing shape of a fleet of wind farms. In a power system, the gap between varying customer demand and rectangular-shaped baseload supply needs to be filled with more flexible, less utilised, hence often more expensive, power generators – just like the gap between varying customer demand and varying solar PV and wind supply needs to be filled.

What's more, the entire concept of baseload, where base-supply power generators are seen as the workhorses and bulk energy suppliers in the power system, is already changing and will become less and less relevant as wind and solar PV begin to form the basis of power supply. The rest of the power system will be optimised around these two weather-dispatched technologies. Wind and solar PV will reduce the quantity of power produced from fossil-fuel plants, along with their capacity factors and operating hours.[23] Electricity generation from controllable, or dispatchable, plants will have to be ramped up and down over relatively short periods to compensate for variable patterns of customers' load on the demand side and of generation from wind and solar on the supply side. In short, all remaining fossil-fuel power stations, both legacy power stations and new-build plants, will need to operate on a flexible basis. So-called baseload power plants that operate with constant output around the clock are counter-productive in such a

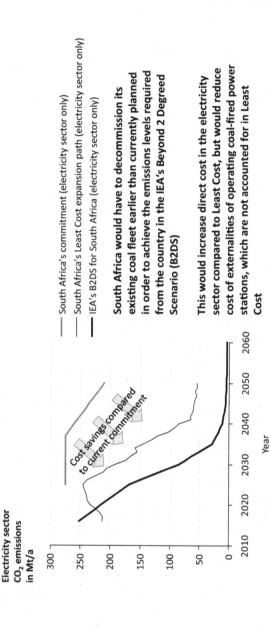

Electricity sector CO₂ emissions in Mt/a

South Africa's commitment (electricity sector only)

South Africa's Least Cost expansion path (electricity sector only)

IEA's B2DS for South Africa (electricity sector only)

South Africa would have to decommission its existing coal fleet earlier than currently planned in order to achieve the emissions levels required from the country in the IEA's Beyond 2 Degreed Scenario (B2DS)

This would increase direct cost in the electricity sector compared to Least Cost, but would reduce cost of externalities of operating coal-fired power stations, which are not accounted for in Least Cost

Figure 3.12 South Africa's current commitment for decarbonising its electricity sector from 2025 to 2050, compared to the CO₂ emissions of the electricity sector if a least-cost pathway is chosen forward, and compared to the emission trajectory that is required from South Africa if the world wants to stay below 2 degrees average global warming (as per IEA's analysis)

power system. They clog the grid with electricity when it is not needed and they are not flexible enough, both technically and economically, to ramp up and down to fill in the gaps between demand and solar/wind supply.

In an article titled 'The Grid Needs a Symphony, Not a Shouting Match', Mark Dyson and Amory Lovins of the Rocky Mountain Institute outlined the position plainly: 'Today, the grid needs flexibility from diverse resources, not baseload power plants'.[24] Dyson and Lovins note, too, that America's national laboratories have already consistently shown that grids with 30% to 80% shares of variable renewable energy are as reliable as fossil fuel-based power systems. In fact, utilities in the US and Europe have had at least a decade of experience in operating grids with declining shares of baseload power relative to renewable energy, and doing so comfortably. A point reinforced by National Grid CEO Steve Holliday, who describes as 'outdated', the idea that large coal-fired or nuclear stations are required for baseload power[25] – National Grid operates the gas and power transmission networks in the UK.

Hence, the importance of flexibility cannot be overemphasised. Flexibility has increased and will continue to increase as wind and solar PV emerge as the default technology choice for low-cost bulk energy in the electricity sector. Nevertheless, it is important for policymakers to assess whether there are indeed enough flexible options available and whether these are cost competitive. In answering this question, much public attention is currently being given to battery storage. This can also be seen from the outcomes of the CSIR's least-cost expansion path, which plans for 8 GW of stationary battery storage by 2030 (refer to Figure 3.6).

However, in the least-cost expansion scenario the assumption is made that the existing fleet of power stations will never be more flexible than what it is today, and that on the demand side only residential warm-water provision and electric vehicles' charging pattern can be made flexible. While stationary batteries are likely to play an increasingly important role in offering flexibility and network stability, they are by no means the only remedy. Flexibility can be met through various supply- and demand-side options and there is already no shortage of solutions.[26]

Firstly, legacy power plants that are readily controllable from a technological perspective can be reconfigured to supplement generation from wind and PV plants. Coal- and gas-fired power plants can be made more flexible through technical and organisational modifications, such as by reducing minimum output rates, increasing load gradients and shortening start-up times.[27]

Secondly, load management and load shifting of existing electricity demand could offer another cost-effective flexibility option. This is especially the case within industry, where it is possible to shift demand by several hours, or where storage capacity for intermediate products, such as heat in the form of steam, cooling in the form of ice, or for compressed air, can be installed. Electrical loads of pumps, for example in the agricultural and water-supply sector, are generally straightforward to make flexible, because of the inherent storage in the form of reservoirs that many pumping processes exhibit. In addition, using smart metres could leverage even small-scale household flexibility options, where these can be shown to be economically feasible. This could take the form of the provision of residential warm water as has been shown to be a valuable intra-day load shifting option in the least-cost expansion path.

Thirdly, all new power generators that are built to accompany solar PV and wind should be flexible from the start. Flexibility means technical flexibility to be able to ramp up and down quickly, to have short start-up and minimum-downtime periods, and to be able to operate at low levels of output in a stable manner. But it also means economic flexibility. Economic flexibility responds to the fact that the power generator providing flexible power will produce less energy per installed capacity than a baseload power generator. Its average annual utilisation (or capacity factor) will be low. That directly leads to those power generators accompanying solar PV and wind that are capital-light and fuel-intensive in their cost structure. If such investments are used seldom, their unit cost does not increase much, as has been shown in the previous chapters. The technology choices are gas-fired power stations, which are cheaper to build than coal or nuclear, but more expensive in terms of fuel in South Africa. It can also mean stationary battery storage, once the investment costs are low enough.

Fourthly, the targeted creation of new, flexible electricity demand, or sector coupling, such as the following:

- Power-to-heat: As shares of electricity generated by wind and solar rises, it could make economic sense to use generation peaks for producing heat instead of burning coal, gas or liquid fuels. Heat production can take place using electrical heating rods in warm water accumulators, or by using heat pumps and, because of the inherent inertia of all thermal processes, can be stored relatively easily. This is not unknown to South Africa, where in the 1990s Eskom incentivised many industrial customers to shift their primary energy consumption away from coal or gas to electricity, because at the time Eskom had abundant coal-fired electricity-generation potential.

- Power-to-pumping: A second source of new flexible demand could be large-scale seawater desalination plants, which require a lot of electric energy. The electricity demand mainly stems from large pumps that are driven by electric motors. Those could be designed to operate in a flexible manner, absorbing the volatility from solar PV and wind supply.

- Power-to-electric-vehicles: As has been shown in the least-cost expansion path, an uptake of electric vehicles can provide a large source of flexible electricity demand. It is also a way of coupling electricity with the transport sector. The opportunities that come with electrification of transport will be further highlighted in the following chapter.
- Power-to-hydrogen: This is another promising deep-decarbonisation and energy-sector-coupling strategy, and we will highlight the opportunities of low-cost, renewables-based hydrogen production for South Africa in the following chapter.

All the flexibility options canvassed here are already technically available and could be implemented at a relatively low cost. The major challenge is arguably not technical implementation, but about providing effective incentives to build infrastructure that is seldom used. Some of the peaking power plants of the least-cost path will only operate a few hours per year. That is most economical from an overall system perspective, but owners need to receive the correct incentives to make such investments.

The implications for South Africa's energy future

So what does this all mean for South Africa's future electricity supply industry? The position probably best summarised by the frequently used (and in this case, misused) 1973 quote by then Saudi Arabian Oil Minister Ahmed Zaki Yamani who argued that the 'Stone Age didn't end for lack of stones'. It is a fact that South Africa's coal resources remain abundant and would be sufficient to meet its future energy needs for many decades to come. However, for environmental and, increasingly, economic reasons these resources are unlikely to remain the foundation of South Africa's electricity and wider energy system.

The future least-cost electricity system will be built on the country's world class solar and wind resources and the country's coal-based legacy system is likely to simply die a natural death. That should not translate into premature plant closures, however. Indeed, South Africa should 'sweat' its coal assets in the interests of ensuring that power prices are controlled as much as possible during the transition, and to provide flexibility rather than the most amount of electricity units. It would be imprudent, for instance, to adopt a stance on coal similar to the one being adopted in Germany regarding the early retirement of the nuclear fleet – a decision that is far from being a least-cost one. Nevertheless, the transition to a renewables-led system cannot be constrained by old notions of baseload, which will become less and less relevant in the years ahead. South Africa should, instead, take full advantage of its comparative wind and solar abundance and use them to transition to a least-cost generation mix, where flexibility is embraced rather than feared.

Bottom Line: South Africa's combined solar and wind resources are world class and it is, thus, technically possible to turn these resources into the workhorses of the country's electricity system. The steep decline in the cost of solar PV and onshore wind has also made these generation sources significantly cheaper than either new coal or new nuclear, even after accounting for the additional costs associated with filling the supply gaps that will arise when the sun goes down or the wind stops blowing. It is therefore least cost to gradually transition South Africa's power system into one that is dominated by solar PV and wind as the new electricity-generation workhorses.

Notes

1 Davenport, J. *Digging Deep: A History of Mining in South Africa*, Johannesburg: Jonathan Ball Publishers, 2013.
2 Cape Monitor. 'Our Mineral Wealth', 7 January 1854.
3 Davenport, J. *Digging Deep: A History of Mining in South Africa*, Johannesburg: Jonathan Ball Publishers, 2013.
4 Chamber of Mines website. *Coal Facts and Figures*.
5 Chamber of Mines. *Facts and Figures 2016*, Published June 2017.
6 Department of Energy. *Actual Tariffs Arising from Department of Energy's Renewable Energy Independent Power Producer Procurement Programme and Coal Baseload Independent Power Producer Procurement Programme*.
7 *Public Presentation Results of Wind and Solar PV Resource Aggregation Study for South Africa*, Pretoria, March 2016.
8 *Wind and Solar Resource Aggregation Study*, Fraunhofer IWES, Eskom, SANEDI and CSIR.
9 *Public Presentation Results of Wind and Solar PV Resource Aggregation Study for South Africa*, Pretoria, March 2016.
10 Bernhard Ernst (NREL): *Analysis of Wind Power Ancillary Services Characteristics with German 250-MW Wind Data*. http://citeseerx.ist.psu.edu/viewdoc/download?doi=1 0.1.1.496.8733&rep=rep1&type=pdf.
11 CSIR. *Statistics of Utility-Scale Solar PV, Wind and CSP in South Africa in 2017*. www.csir.co.za/csir-energy-centre-documents.
12 CSIR Website, accessed November 2017, A Normalised Power Feed-In For Wind www.csir.co.za/documents/supply-area-nomalised-power-feed-windxlsx.
13 *Public Presentation Results of Wind and Solar PV Resource Aggregation Study for South Africa*, Pretoria, March 2016.
14 Fraunhofer IWES, CSIR, SANEDI, Eskom: *Wind and Solar PV Resource Aggregation Study for South Africa*, page 47. www.csir.co.za/sites/default/files/Documents/Wind_and_PV_Aggregation_study_final_presentation_REV1.pdf.

15 Wright: *Long-Term Electricity Sector Expansion Planning Outcomes: A Unique Opportunity for a Least Cost Energy Transition in South Africa.* www. researchgate.net/publication/320871338_Long-term_electricity_sector_ expansion_planning_outcomes_A_unique_opportunity_for_a_least_cost_ energy_transition_in_South_Africa.
16 Bischof-Niemz, T. *The Case for Renewable Energy to Provide Base Load Energy in South Africa.* Presentation to Presentation at the Sustainability Week 2017, June 2017.
17 *South African Department of Energy, Market Overview and Current Levels of Renewable Energy Deployment,* pages 27 and 28. www.energy.gov.za/files/ renewable-energy-status-report/Market-Overview-and-Current-Levels-of Renewable-Energy-Deployment-NERSA.pdf.
18 The value of wind revisited: A systems-planning perspective, Jarrad Wright, CSIR, https://www.strommarkttreffen.org/2017-10_Wright_Value_of_wind_ revisited.pdf
19 CSIR: *Future Wind Deployment Scenarios for South Africa.* www.researchgate. net/publication/321098174_Future_wind_deployment_scenarios_for_South_ Africa.
20 South African Department of Energy, Draft Integrated Resource Plan 2016, http://www.energy.gov.za/IRP/2016/Draft-IRP-2016-Assumptions-Base-Case- and-Observations-Revision1.pdf.
21 Steyn, G., Burton, J. and Steenkamp, M. 'Eskom's Financial Crisis and the Viability of Coal-Fired Power in South Africa: Implications for Kusile and the Older Coal-Fired Power Stations'. *Meridian Economics,* Cape Town: South Africa. http://meridianeconomics.co.za/wp-content/uploads/2017/11/CoalGen-Report_ FinalDoc_ForUpload-1.pdf, 2017.
22 *International Energy Agency: Energy Technology Perspectives 2017.* www.iea.org/ etp2017/.
23 Agora Energiewende, '12 Insights on Germany's Energiewende', February 2013. www.agora-energiewende.de/en/die-energiewende/12-insights-of-germanys energiewende/.
24 Dyson, M. and Lovins, A. *Rocky Mountain Institute.* The Grid Needs a Symphony, Not a Shouting Match, June 2017. www.rmi.org/news/grid-needs symphony-not-shouting-match/, www.aiche.org/chenected/2016/03/chinese-grid officials-explode-myth-baseload-power-ceraweek.
25 Beckman, K. *Energy Post. National Grid CEO: Large Power Stations For Baseload Power Is Outdated,* September 2015. http://energypost.eu/interview-steve-holliday ceo-national-grid-idea-large-power-stations-baseload-power-outdated/.
26 Agora Energiewende, '12 Insights on Germany's Energiewende', February 2013.
27 International Energy Agency, Status of Power System Transformation 2018, Advanced Power Plant Flexibility, https://webstore.iea.org/status-of-power- system-transformation-2018.

4 Electrification of just about everything

The industrial revolution, Andrew McAfee and Erik Brynjolfsson write in their 2014 *New York Times Bestseller* 'The Second Machine Age', is not a story of steam power alone, but steam was the very start of it all.

> More than anything else, it allowed us to overcome the limitations of muscle power, human and animal, and generate massive amounts of useful energy at will. This led to factories and mass production, to railways and mass transportation. It led, in other words, to modern life,[1]

the two Massachusetts Institute of Technology academics elucidate.

What is often overlooked however is the fact that all thermal processes that are at the core of our industrialised world have a significant downside: the inherently low efficiency of the Carnot cycle. Most of the energy that is stored in our primary energy today (mostly fossil fuels) is wasted in the form of heat when the fossil fuel is burned, and only small portions are brought to actual end-use (driving a turbine, a motor, etc.). Secondly, the three dominant fossil fuels (coal, natural gas and oil) are all supplying distinctly different end-use sectors (power, power and heat, and transport), and therefore little portfolio effect between these sectors is reaped.

Turning electricity into the new primary energy source has significant upsides: most loss-heavy heat cycles can be eliminated by making use of wind and solar PV electricity, and secondly, electricity as new primary energy brings together previously separated end-use sectors, allowing for portfolio effects across the entire energy sector to kick in. In this chapter, the case is therefore made for why South Africa should pursue an electrification-of-everything strategy; coupling its progressively decarbonised and least-cost electricity system, as outlined in Chapter 3, to the other two energy end-use sectors: transportation and heating/cooling.

There are already strong signs that other countries are starting to accept that electricity will play a larger role in their future energy systems, with the International Energy Agency's (IEA's) most recent modelling exercises presenting a scenario whereby electricity comprises 40% of the rise in final consumption to 2040 – the same share of growth that oil took for the last 25 years.[2] Electricity, the IEA states, is becoming the energy of choice in most end-uses, driven by several factors, including the adoption of electric vehicles (EVs), both battery and fuel-cell EVs (BEVs and FCEVs), and heat pumps, which will result in passenger-vehicle and heating-energy demand increasingly supplied by electricity. Moreover, it is anticipated that industrial production processes will require more electricity, as will millions of new middle-income families in developing countries, who will add appliances and install cooling. The IEA's modelling shows that 74% of the electricity demand growth will come from industrial motor systems, space cooling, appliances and ICT. Overall, the scenario is for total electricity demand to grow at 2% a year from 2016 to 2040, which is nearly twice the rate of growth of final energy demand.

That implies that electricity will experience more growth than all other fuels, meeting over 37% of additional final energy demand.[3] In its New Policies Scenario, the IEA sees electricity demand rising by 60% to 2040, with over 85% of global growth occurring in developing economies. As a result, total investment in the power sector to 2040 is modelled at $19.3 trillion, representing almost half of total energy-supply investment. The IEA's most recent scenarios also present the case for this rising power demand to be met, increasingly, from solar photovoltaic (PV) and wind generation technologies, with renewables capturing two-thirds of the investment in power plants. Even though electricity supply expands by almost 60% to 2040, the IEA scenario is for related carbon dioxide (CO_2) emissions to rise by just 4%, as global average emissions intensity of power generation drops. Nevertheless, it cautions that, even though CO_2 emissions peak before 2040, the current decarbonisation trajectory still sets the world on course for a global mean temperature rise of roughly 2.7°C by 2100 and hence significantly higher than the binding Paris Agreement, of 2015, where 174 States and the European Union agreed to 'holding the increase in the global average temperature to well below 2°C above pre-industrial levels and pursuing efforts to limit the temperature increase to 1.5°C above pre-industrial levels, recognizing that this would significantly reduce the risks and impacts of climate change'.[4]

These powerful trends towards renewables and the electrification of energy services have significant implications for South Africa, which is no foreigner to efforts to accelerate the conversion of households, factories and farms from other energy sources to electricity. Indeed, prior to the recent supply squeeze and the associated tariff hikes, South Africa's national utility had active, and successful, conversion programmes. Looking to the future, the principle of higher electrification on the back of cheap power remains intact; all that has changed is the least-cost primary electricity mix. Instead of coal, the country's comparatively superior electricity source is now solar and wind, which will allow it to produce electricity more cheaply than most other countries. The potency of this advantage rises even further in a context where the world is increasingly turning to electricity, through sector coupling, as a source of energy for industry, mobility and heating. Given its cheap and decarbonised power advantage, South Africa has the opportunity to further entrench and expand the electricity intensity of its energy system in a way that makes it an investment destination of choice for electricity-intensive manufacturing, as well as the production of green, electricity-based chemicals and fuels.

Primary energy reduction

Before elaborating on this potential, however, it is important to highlight a key characteristic of an electricity-led energy model. As the energy system transitions to one where electricity is the backbone and where solar and wind are the power-generation workhorses, the amount of primary energy falls. This is true even when electricity consumption expands to displace other fuel sources in the transport and heat sectors. Such a counterintuitive outcome arises as a result of the efficiency of the new electricity generators, relative to conventional fossil-fuel plants, as well as the efficiency of the emerging transport and heating solutions. It essentially relates to the inefficiency of the Carnot cycle,[5] which humanity has used for two centuries to convert thermal energy, by burning fossil fuels, into motion in either turbines to generate power or in piston-stroke internal combustion engines to generate propulsion.

Take a coal-fired condensing power station, for instance. Such plants are 30% to 40% efficient, which means that for every unit of electricity produced, about three units of primary energy, in the form of coal heat value, is required. Some 60% to 70% of the energy content of that primary energy coal is lost through the smoke stack or through the cooling towers to unusable heat. In the transport sector, the inefficiency of the Carnot cycle is even more intriguing: an internal-combustion-engine-propelled vehicle is only around 20% to 25% efficient, with the efficiency especially poor when a car is used in stop-start urban driving. That means up to 80% of the primary energy put into an urban passenger car in the form of either diesel or petrol is not actually used for propulsion, but is wasted in the vehicle's radiator, through its exhaust pipe, or in its brakes in the form of heat.

On the side of heating, the picture looks a bit rosier, because the final end-use energy form required (heat) is actually the same as what would usually be considered a waste or a loss. A gas-fired condensing boiler converts up to 90% of the heat content of the natural gas into useful space heat, while electric resistive heaters essentially convert all electricity going into them into space heat.

But still, in all three cases (power generation, transportation and heating) new technologies exist now with much higher efficiency levels than what we were used to.

Wind and solar photovoltaic (PV) generation do not exhibit the conversion losses (heat) related to the combustion process of conventional fossil-fuel- or uranium-based power generation (refer to definitions by IEA and IPCC[6]). Both technologies produce electricity directly, without first having to produce steam to drive a generator. Electricity from a wind turbine is produced when wind turns blades that spin a shaft connected to a generator, which makes electricity. Solar PV panels, meanwhile, convert the sun's rays into electricity by using photons, or light particles, to knock electrons free from atoms in silicon cells. When connected to a circuit, these electrons generate a flow of electricity. Without these heat-related losses, the definition of primary energy for both solar PV and wind is exactly the same as the amount of electricity they produce. As a result, statistically, around two-thirds less primary energy is required to produce the same amount of electricity as would be the case with a traditional fossil-fuel plant.

On the transportation side, a BEV is about 80% efficient, for two reasons: firstly, it does not use a combustion engine to produce propulsion in the first place. Instead, it makes use of a highly efficient battery-inverter-electric-motor drivetrain. Secondly, because it is electrically driven, it can recuperate part of the kinetic energy when braking and recharge its batteries. Recovering some of the initial fuel used is not possible for a diesel or petrol car. For such vehicles, every brake manoeuvre irreparably destroys kinetic energy that was already inefficiently produced by burning a fossil fuel in the first place.

For heating and cooling, heat pumps have made their way into the market for mainstream energy devices. A heat pump does exactly what it says it does, it pumps heat from one space to another – from outside to inside (space heating) or from inside to outside (space cooling or a refrigerator). It does so by using electricity as the driving force for the pump. However, the pump requires far less energy than the amount of heat content that it pumps. In this way, every kilowatt-hour of electricity is 'leveraged' and can pump up to three to four units of heat.

Figure 4.1 conceptually summarises this large-scale reduction of primary energy demand that arises when moving to a renewables-based, electric-vehicle and heat-pump-driven energy system, from today's fossil-fuel-based, Carnot-cycle-driven system, with its inherent inefficiencies. It is important to note that the reduction of primary energy is not associated with a reduction in end-user utility from energy services. The energy services stay exactly the same.

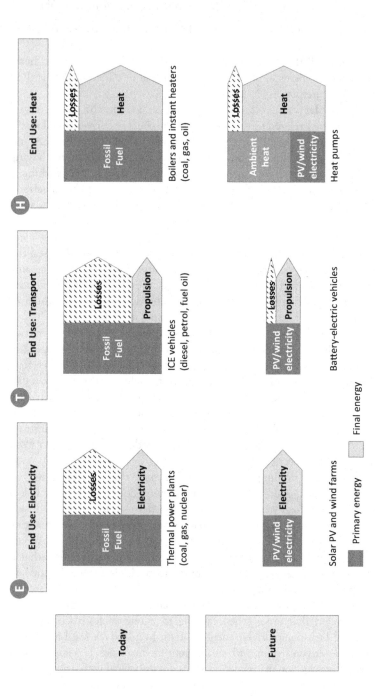

Figure 4.1 Primary energy consumption versus useful final energy and associated heat losses in today's and tomorrow's energy system[7]

South Africa's energy system today: domestic coal and imported oil

Having understood the counterintuitive primary energy dynamic outlined previously and having accepted the technical and economical feasibility of turning wind and solar into the country's electricity generation work-horses, the logical next question is what it means for the South African energy system as a whole. Currently, all three energy services – electricity, transportation and heat – are met primarily through domestic coal and imported oil, with relatively minor contributions in the electricity sector from nuclear and renewable energy, and from biomass in the heating sector. As discussed previously, coal plays a disproportionately larger role in South Africa's energy mix than in many other countries, owing to both Eskom's almost exclusive coal-exposure and Sasol's conversion of the energy mineral into liquid fuels, using the Fischer-Tropsch process.

In 2015, 6 150 petajoules (PJ) of primary energy from coal were produced in South Africa. Of that, roughly a third, or 2 115 PJ, were exported to countries that burn South African thermal coal to produce electricity. A total of 2 582 PJ were used to produce 200 TWh of electricity (equal to 720 PJ), while 217 PJ of coal, together with 105 PJ from natural gas, were deployed in the liquefaction process to produce 213 PJ of synthetic liquid fuels (roughly 7–8 billion litres or 30% of South Africa's petrol and diesel demand). The primary energy provided by uranium into the electricity sector was 134 PJ, while the renewables contribution is 684 PJ, which is mostly wooden biomass deployed for producing heat in industrial processes and for residential space heating.

The South African transport sector consumed 838 PJ in 2015. Roughly 70% of the primary energy for this sector is derived from oil imports, and 30% is supplied by coal and natural gas converted into transportation fuels by Sasol and by PetroSA.

At present, there is limited sector coupling, as well as limited opportunity to decarbonise the energy economy by integrating electricity with transport and heat, owing to the predominance of coal. However, should South Africa progressively decarbonise its electricity generation over the coming three decades on the back of its formidable solar and wind resources, it is going to make increasing sense to assess whether it is economically and technically feasible to extend, through coupling, the electricity sector's new-found decarbonisation to the areas of transportation and heat.

We can also see that the South African energy system in the real world exhibits the inefficiencies that were conceptually outlined in Figure 4.1. These inefficiencies can be attributed to the country's reliance on the

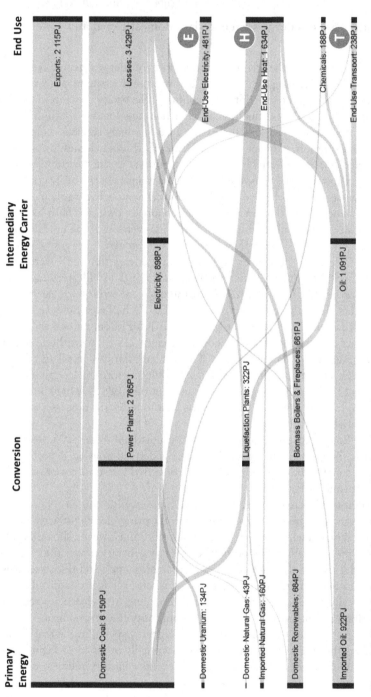

Figure 4.2 Simplified energy-flow diagram (Sankey diagram) for South Africa in 2015[8]

Carnot cycle to convert heat stored in fossil fuels into useful work by burning them. Of roughly 6 000 PJ of total primary energy supplied to the South African economy each year (total production plus imports minus exports), roughly 3 430 PJ were wasted as heat during conversion and transmission steps, and only 2 350 PJ were brought to actual end-user energy services, and another 190 PJ are used as non-energy in the chemicals sector (see Figure 4.2). It is important to keep in mind that it is the final end-use energy, rather than the primary energy, that is of ultimate interest, because it is this energy that provides the energy services. The coal-dominated primary energy supply into the South African economy leads to roughly 450 million tonnes of CO_2 emissions per year.[9]

Sector coupling

To recap, in Chapter 3 we showed how a large country such as South Africa, with its almost unrivalled solar and wind resources, faces no constraint in using solar PV and onshore wind (now the cheapest sources of new generation) to supply the bulk of its electricity demand. These two workhorses would need to be supported by flexible generation, initially the coal fleet and load-following fossil-fuel plants. Over time, though, this flexibility could be derived from decarbonised solutions, including: dispatchable renewables, including: biogas from municipal, agricultural, water and animal waste; biomass generators; hydropower; concentrated solar power plants and power-to-power storage, including pumped hydro and batteries.

The question then arises as to whether it is technically feasible and economically viable to couple South Africa's least-cost, decarbonised electricity system to the energy service requirements of transportation and heating. Left to operate in their traditional sectoral and subsectoral silos, such an outcome looks challenging to say the least, particularly in the area of transportation. The picture changes significantly, though, when a more joined-up view is taken. Such perspectives will also not be unique to South Africa, with the IEA's modelling already signalling that electricity will play an increasing role in the provision of energy services in industry and transport. On the supply side, the agency lists South Africa together with Australia, the Horn of Africa, North Africa, northern Chile, southern Peru and Patagonia, as well as several regions in China and the Midwestern US, as having particularly abundant and cheap renewable resources. These resources could be used not only to drive a fleet of BEVs and provide residential and commercial heat for water and cooking, but also to produce green hydrogen-rich fuels and chemicals. We assess some of these sector-coupling opportunities in the pages that follow.

Transport services

South Africa's existing ground-based, aviation and maritime transport services for moving people and goods are enabled and energised using the primary energy sources of imported oil and domestic coal. Some 26.7 billion litres of liquid fuels are currently consumed yearly, comprising 11 billion litres of petrol, 13 billion litres of diesel, 2.2 billion litres of jet fuel and 500 million litres of fuel oil.[10] Therefore, a natural first place to look for a decarbonised solution is whether so-called drop-in fuels, derived from energy crops rather than fossil fuels, can meet overall demand. In other words, can liquid biofuels such as biodiesel and bioethanol be produced at the scale required to fuel the country's passenger cars, freight trucks, diesel locomotives, planes and ships?

South Africa has 120 000 km² of arable land.[11] At an estimated yield of biofuels of 2 000 litres per hectare per year, even using all the arable land available, which would obviously not be justifiable from a food-security perspective, South Africa could, theoretically, produce only 24-billion litres of biofuels yearly, which is below the prevailing annual demand and could, thus, not accommodate further growth. Liquid biofuels, therefore, have limited technical potential to decarbonise the transport sector. However, they could well play a supplementary role, particularly in aviation, where demand is more moderate (2.2 billion litres) and where immediate prospects for electrification are low. Nevertheless, apart from niche sectors, biofuels cannot be relied upon to decarbonise the South African energy system as a whole.

The next area to interrogate is whether a decarbonised outcome is possible by coupling South Africa's future renewables-led electricity system to the transport subsectors of road, rail, air and sea. When one does that, two interesting possibilities arise: power to e-mobility, through BEVs and FCEVs; as well as power-to-liquids, partly using the platform already laid by South Africa's investment into the large-scale use of the Fischer-Tropsch process to convert coal to fuel and chemicals.

In the area of e-mobility, the case for replacing liquid fuels with electricity for ground-based transportation is compelling, with South Africa having a number of comparative advantages in its urban areas that are supportive of an accelerated mass adoption of BEVs. In addition, the country's road-to-rail freight strategy could further stimulate the shift to greater electrification of rail transportation. In the longer term, there is also potential to electrify the country's road haulier fleet, but this would require new infrastructure investment. The same is not true, though, for aviation and shipping, where the decarbonised solutions lie elsewhere. As mentioned previously, there is some potential in aviation to pursue decarbonisation through the introduction of liquid biofuels. However, for both shipping and aviation another part of the solution may lie in South Africa's power-to-liquids potential, which we explore later.

In the area of passenger vehicles, the opportunity being created by the global trend among automakers to transition to BEVs, combined with some geographic and other idiosyncrasies in the way urban South Africans use cars, could translate into a faster-than-anticipated adoption of BEVs in South Africa. And, as long as the electricity being consumed when charging the BEV is premised on a renewables-heavy electricity mix, the decarbonisation potential is potent.

A global transition to EVs is undoubtedly under way, with just about every automobile manufacturer having outlined how they see EVs fitting into their future production and sales. In its Electric Vehicle Outlook 2017, Bloomberg New Energy Finance (BNEF) forecasts that plunging lithium-ion battery prices, which are predicted to fall more than 70% by 2030, will accelerate the uptake of EVs in the second half of the 2020s. BNEF has, therefore, revised its previous estimate that EVs will account for 35% of new car sales by 2040 to 54%. 'By 2040, EVs will displace eight-million barrels of transport fuel per day, and add 5% to global electricity consumption'.[12]

There are some peculiar features about South Africa's urban landscape, though, that could further stimulate EV adoption and penetration. Firstly, the minibus taxi industry, which facilitates around 15 million commuter trips daily,[13] is arguably an ideal e-mobility match. For one, such taxis typically operate within the 200- to 300-km daily range already provided by most EVs. Ironically, South Africa's lingering apartheid geography, which sees many taxi users endure long but predictable daily commutes, aligns with that range profile. In addition, while EVs are expensive to buy, their fuel costs are lower than traditional internal combustion engine (ICE) vehicles. Here again, the consistently high utilisation levels of minibus taxis will allow that cost premium to be recovered through the relatively low EV fuel costs. Indeed, on a cost-per-kilometre basis, an EV taxi truly comes into its own when it is used to its maximum range every day, which would be the case with most minibus taxis. Moreover, peak taxi usage will generally overlap with peak electricity demand. As a result, vehicle charging is unlikely to coincide with evening and morning power peaks. Lastly, taxi routes are also regulated and predictable, which will help with the planning and deployment of charging-infrastructure.

The South African National Taxi Council estimates that the current taxi fleet is made up of some 130 000 vehicles, of which about 95 000 vehicles are used for short- and medium-distance trips in urban centres. In fact, more than one-third of the vehicles operate in the highly populated Gauteng province.[14] Electrifying this privately owned fleet would be a major undertaking and would involve large investments. Coordination between the industry, government, the banks and the electricity system operator would also be essential. Nevertheless, it presents a very real opportunity to decarbonise public transport in South Africa, while improving business sustainability by lowering operating costs for taxi owners. The same logic would apply for the bus and rail passenger transport network, which are currently responsible for about 9 million and 2 million daily commuter trips, respectively.[15]

Besides the potential to decarbonise road transportation through coupling BEVs to a renewables-led generation mix, there could be profound foreign exchange and industrialisation benefits for South Africa. A comparison of the total cost of ownership of an ICE car and an equivalent BEV, which travels 25 000 km/a over an average daily commute of 100 km, shows that BEVs are already cheaper. Assuming a fuel economy for the ICE of 8 l/100 km against the BEV's 15 kWh/100 km, the fuel consumption stands at 2 000 l/a for the ICE and 3 750 kWh/a for the BEV. At R14/l and R2/kWh, the yearly fuel cost of the ICE is R28 000 as opposed to the BEV's R7 500. Should the BEV cost R400 000 to buy against an ICE at R300 000, the yearly instalments for the BEV would be higher, at R65 000, against R49 000 for the ICE. However, maintenance on the BEV will be lower, owing to the fact that such vehicles have fewer moving parts. A BEV has no clutch, no gearbox, no high-maintenance combustion engine, and the brake discs and brake pads are barely used, because, under normal deceleration, the BEV is slowed down by recuperating energy back into the batteries, i.e. running the electric motors as generators. If we assume that insurance and other non-technical fixed costs would be the same for BEV and ICE, the total fixed cost would stand at R15 000 a year for a BEV, against R20 000 a year for the ICE. Therefore, the yearly total cost of ownership is R87 500 for the BEV (or R3.5 per km), as opposed to R97 000 for the ICE (or R3.9 per km) (see Table 4.1).

From a trade balance perspective, fuelling the BEV with domestically produced electricity lowers South Africa's cash outflow for transportation fuel to zero, as compared with $1 000 for an ICE with a fuel consumption of 2 000 l/a. Hence, 1 million EVs, replacing 1 million ICEs, would save the country $1 billion or R12 to 13 billion per year of cash that is spent every year purchasing fuel from abroad.

Because of the high efficiency of BEV transportation, the million-vehicle-strong BEV fleet would increase electricity demand by only 3.75 TWh/a, or 1.7%. However, this new demand could benefit South Africa's national utility, Eskom, in two ways. Firstly, it would increase electricity sales and, secondly, the utility could use the BEV fleet to shift load into the off-peak periods, especially overnight, when BEV cars would typically be charged. This would lead to higher loading of the coal fleet during these off-peak hours, which will prove beneficial over the next decade or

Table 4.1 Total cost of ownership for a conventional car with internal combustion engine (ICE) versus a battery-electric vehicle (BEV) with battery and electric motor drivetrain

Parameter	ICE	BEV
Daily commuting	100 km/d	100 km/d
Workdays per year	250 d/a	250 d/a
Distance travelled	**25 000 km/a**	**25 000 km/a**
Fuel economy	8 l/(100 km)	15 kWh/(100 km)
Fuel consumption	2 000 l/a	3 750 kWh/a
Fuel price	14 R/l	2 R/kWh
Fuel cost	**28 000 R/a**	**7 500 R/a**
CAPEX	300 000 R	400 000 R
Instalments (10 a, 10%)	**49 000 R/a**	**65 000 R/a**
Insurance, license	10 000 R/a	10 000 R/a
Maintenance service	10 000 R/a	5 000 R/a
Fixed operations costs	**20 000 R/a**	**15 000 R/a**
Total cost of ownership	**97 000 R/a**	**87 500 R/a**
... per km	3.9 R/km	3.5 R/km

so, where the existing coal fleet and its cheap marginal cost of production is still required to accommodate increasing levels of variable solar PV and wind power.

In addition, South Africa's urban geography, as well as certain characteristics of the South African household, could well prove equally supportive of BEV adoption by private car owners. Take Gauteng, the city-region where more than 13 million of South Africa's 55 million citizens reside. It is the country's smallest province by area, with a radius from its centre to its fringes never exceeding 100 km. Even so, a number of private car owners drive 100 to 120 km to work and back daily; the Pretoria-to-Johannesburg commute. As with the minibus taxi, this relatively long and consistent commuter profile is a good match for BEVs that have a charging range of around 300 km. Fuel switching will, therefore, bring the same cost-per-kilometre advantages. More importantly, though, is the fact that most car owners in Gauteng have a dedicated under-cover parking spot at home. This reality offers an immediate charging-infrastructure advantage over Europe, for example, where car owners typically park on the street. Put differently, having a dedicated parking spot makes it far easier for a Gauteng resident to access either a plug point at home, or to make the decision to invest in a permanent fast charger. Even if a BEV owner doesn't install a fast charger, the car can still be charged from empty to full, on a normal power socket, in roughly eight hours – or overnight. Conditions in Gauteng are thus, arguably, almost ideal for high BEV adoption.

Power to gas and liquids

Apart from e-mobility, there is further potential in South Africa to decarbonise all three transportation modes (land, air and sea) through power-to-gas and power-to-liquids solutions. Moreover, this could be done on a cost-competitive basis by using the platform already created by Sasol's investment into the Fischer-Tropsch technology that currently employs coal and natural gas to manufacture liquid fuels and chemicals. The idea is to produce so-called green hydrogen and hydrocarbons, such as methane, methanol and other 'drop-in' liquid fuels, by coupling Sasol's production network to South Africa's increasingly renewables-led electricity industry. To make such a radical change, coal and gas would be progressively displaced by renewable electricity, water and carbon dioxide (CO_2) to produce these hydrocarbons (see Figure 4.3).

Although South Africa is a water-constrained country, the amount of water required to support such a process would not be a limitation, particularly as the electricity system transitions increasingly to generation technologies that do not require water. Currently, Eskom consumes more than 300 billion litres of fresh water annually. Theoretically, to produce the entire 26.7 billion litres of liquid fuels currently consumed, about 100 billion litres of fresh water would have to be used, which represents only one-third of what Eskom currently uses in power generation. Since the power sector itself will reduce its water consumption by more than 90% if the least-cost expansion path towards renewables is followed (refer to Chapter 3), there is still an overall reduction in water demand. In addition, should EVs gain traction in the passenger transport sector, demand for liquid fuels will also decline. The synthetic, electricity-based fuels outlined here would only be used to supply the residual demand for fuels and chemicals that cannot be replaced with electricity directly. The CO_2, meanwhile, could be sourced from various processes: initially from existing fossil CO_2 sources at Sasol; and later from biogas-to-electricity plants, which produce CO_2 as a by-product, as fossil CO_2 is progressively phased out. In the very long run, CO_2 for such processes could be extracted directly from the air.

In addition to being the basis for subsequent hydrocarbons production, the hydrogen produced from water and electricity can also be used as a fuel directly: either in fuel-cell driven electric vehicles, FCEVs, or in the steel-making process for the reduction of iron-ore to iron or in the hot-dip galvanising process. The hydrogen can furthermore be the basis for ammonia production, where ammonia is the key feedstock for fertilisers, which are in high demand globally. In all three cases the resulting fuel (hydrocarbons, hydrogen or ammonia) would be carbon-neutral. The energy content of the fuel is derived from electricity, and, if that electricity is renewable, then

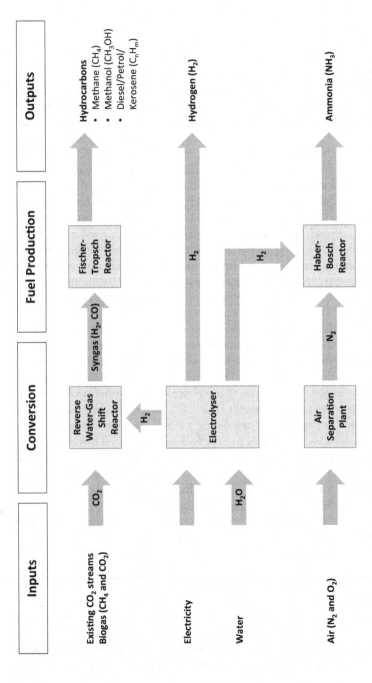

Figure 4.3 Production of synthetic fuels (hydrocarbons, hydrogen and ammonia) from electricity, water, carbon dioxide and air

there is no introduction of fossil carbon into the process. The other feed-stocks of water, air and carbon dioxide are merely carriers of certain atoms that are required at some step during the process, but do not contribute to the overall energy yield. The electricity is the new primary energy source.

To achieve a power-to-fuels outcome, three technological processes are required. Firstly, electrolysis is needed to split water using renewable electricity. Electrolysis of water is the decomposition of water into oxygen and hydrogen by passing an electric current through the water. Already, a small proportion of global hydrogen is produced from electrolysis, usually as a by-product of chlorine production.[16] But the majority of all global hydrogen today is produced from natural gas or from coal (in Sasol's case). Secondly, a reverse water-gas shift reaction is necessary to produce carbon monoxide from the hydrogen and CO_2. Thirdly, a Fischer-Tropsch reactor converts the mix of carbon monoxide and hydrogen (called syngas) into hydrocarbons.

Sasol currently uses water-gas shift reactors to convert water and carbon monoxide, made through coal gasification, to produce hydrogen and CO_2. The so-derived hydrogen and further carbon monoxide is then converted into hydrocarbons in the Fischer-Tropsch process. That means that Sasol already houses two of the essential process steps to produce liquid fuels from electricity. The only infrastructure components absent are the electrolysers to produce the hydrogen, currently derived from natural gas and coal.

In a case where the hydrogen is being produced from electricity through electrolysis, some of the Sasol reactors would need to be reversed to produce carbon monoxide and water. This might be done by injecting a portion of the hydrogen arising from of the electrolysis, together with CO_2, into a reverse water-gas shift reactor. The CO_2, meanwhile, would be sourced from Sasol's existing processes during a transition phase of probably two to three decades. The balance of the hydrogen produced using electrolysis would flow directly into the Fischer-Tropsch process, where it is combined with the carbon monoxide (now produced in the reverse water-gas shift reactors) to make hydrocarbons. The hydrocarbons are any form of synthetic fuels that Sasol already produces today.

This conceptual thinking has already moved beyond pure academics. As Agora Energiewende writes, 'selective use of synthetic fuels is essential to decarbonise German transport, industry and heating'.[17] As can be seen in Table 4.2, synthetic fuels based on renewable electricity are amongst the lowest possible greenhouse gas emissions of all transport fuels. In terms of the well-to-wheel energy requirement, however, this is very high, owing to the associated multiple conversion steps needed before a primary-energy unit at the well eventually arrives at, and drives, the wheels. This means that power-to-liquid fuels will certainly not be used where it is possible to use a BEV or a FCEV. But in situations where it is simply not possible to decarbonise using electric transport, such as in aviation, power-based

Table 4.2 Energy demand and CO_2 emissions for different ways of providing mobility[18]

Fuel category	Fuel/engine combination	Greenhouse gas emissions in gCO_2 eq/km	Well-to-wheel energy consumption in MJ/(100 km)
Electricity	Electricity in BEV (wind)	0	40
	Electricity in BEV (EU mix)	55	120
	Hydrogen in FCEV (wind)	0	100
	Hydrogen in FCEV (reforming of natural gas)	60	110
Fossil fuels	Diesel in hybrid	75	105
	Petrol in hybrid	80	110
	Diesel in ICE	105	140
	Petrol in ICE	125	170
	Natural gas in ICE	100	170
Biofuels	Biodiesel (rapeseed) in ICE	65	255
	Ethanol (sugar beet)	55	350
	Biomethane (maize)	60	330
	Biomethane (municipal waste)	25	290
Synthetic fuels	Synthetic diesel (PtL, renewable electricity)	0	305
	Synthetic methane (PtG, renewable electricity)	0	300

liquids will play a role. The German association of the oil industry (MWV) in a recent study concluded that the import of what they call "e-fuels", electricity-based fuels, from sun- and wind-rich countries will play a crucial role in the German decarbonisation efforts. Even if only a small fraction of the German liquid fuel demand is converted, it is still a massive export opportunity for South Africa.[19]

The IEA, in an insight paper titled 'Renewable Energy for Industry', states that the drastic cost reductions in solar PV and wind open new possibilities for competitive green hydrogen production in large-scale plants.[20] The report outlines how the renewables-based electrolysis of water could produce hydrogen-rich chemicals and fuels, such as ammonia or methanol. These could be used in various industries as precursors, process agents and fuels, as well as in other end-use sectors such as buildings and transport. In regions such as South Africa, where resources are especially abundant, the cost of hydro, solar and wind power, when combined, can fall below $0.03/kWh and supply high load factor electricity demand profiles. Such low electricity prices may allow hydrogen to be produced at costs that are competitive with natural gas reforming, oil-cracking or coal gasification, but without the associated CO_2 emissions. Global CO_2 emissions associated with manufacturing nitrogen-based fertilisers are currently estimated at 420 million tonnes a year.[21]

For the IEA, a potential opportunity for decarbonising industry, on the back of cheap renewables, is ammonia, which is used mostly to manufacture nitrogen fertilisers. In some cases, green ammonia could immediately compete with conventionally produced ammonia. The report calculates the cost of producing green ammonia to be between $400/t and $700/t, depending on the cost of the electricity and the load factor of the electrolyser. By comparison, natural gas-based ammonia production currently ranges in cost from $200/t to $600/t, with the lowest-cost production currently arising in the US, owing to its abundant and inexpensive shale-gas resources. The ammonia could be shipped from the best resource areas at a cost of between $40/t and $60/t, depending on distance and vessel size. This green ammonia could be more affordable than conventionally produced ammonia in some cases, or would require a modest carbon price of approximately $25, mainly to cover transportation costs.

Besides ammonia production, the IEA paper argues that green hydrogen could also help decarbonise industry by serving as a precursor to manufacturing methanol and other chemicals. In addition, renewables-based hydrogen could be used to reduce iron-ore to pig iron, which could then be melted in electric arc furnaces with scrap. Emissions associated with cement manufacturing could be reduced by using solar or electric heat, or by combusting hydrogen-rich synthetic fuels, the report adds. However, applying solar and wind power to current levels of production in the chemicals, iron and steel, and cement sectors would entail several terawatt-hours (TWh) of additional renewable-energy production not already factored into the IEA's long-term low-carbon scenarios. Overall, though, the IEA paper asserts that a combination of direct process electrification, as well as the use of storable hydrogen-rich chemicals and fuels manufactured from electricity, may offer the greatest potential for renewables uptake by various industries.[22]

Methanol, meanwhile, which is a versatile alcohol used in chemicals and fuels, could also be produced using clean electricity. In Iceland, a company known as Carbon Recycling International is producing 4 000 t/a of carbon-neutral methanol, based on the electrolysis of water using renewable electricity from a geothermal power plant and CO_2 captured from co-located industrial processes. However, in the absence of industrial sources of CO_2, methanol production from renewable hydrogen and CO_2 captured from ambient air is also technically feasible. This is, however, the long-term end game for a world in which CO_2 sources do not exist anymore. During a transition phase of one to two decades there will still be an abundance of CO_2 from existing fossil sources – although it is declining.

Leveraging Sasol's existing asset base

In South Africa, which has a rich heritage of producing fuels and chemicals from unconventional sources, the implications could be significant. Sasol, for instance, could progressively begin replacing its fossil hydrogen with green hydrogen to decarbonise part of the liquid fuels and chemicals being produced at Secunda and Sasolburg. Ultimately, though, entirely new processes could be created around this green hydrogen, which would enable South Africa to produce a portfolio of clean fuels, chemicals and fertilisers for domestic and international consumption. There may even be a real potential for turning South Africa into the 'Saudi Arabia' of green fuels and chemicals. This could be achieved by blending its comparative solar and wind advantages with its unique Fischer-Tropsch experience and asset base, to produce decarbonised products for export. However, such an outcome will depend materially on whether the country takes proactive steps to commercialise the electrolysers, which is a key missing link if South Africa is to realise such an outcome. To be sure, not all the prospects will prove commercially viable. Nevertheless, South Africa's research councils, as well as its energy and industrial policymakers, need to become more fully alive to the potential that exists.

In summary, South Africa has two distinct competitive advantages. Firstly, the country's renewable electricity will be cheaper than in most other countries, because of its excellent solar and wind resources, combined with vast amounts of land that is readily available for the deployment of renewables. This competitive advantage is permanent and cannot be taken away. Secondly, South Africa has vast experience in the creation of synthetic liquid fuels, which is unique globally. Sasol is the largest coal-to-liquids producers in the world. This experience, together with the existing asset base of large-scale synfuel production, is a second competitive advantage. However, it is an advantage that could, in principle, be overtaken by innovative follower countries. Nevertheless, it represents a typical first-mover advantage that should be utilised and exploited before others catch up.

This combination provides a huge opportunity for South Africa to commercialise renewable-electricity-based, carbon-neutral synthetic fuels from
power-to-liquid processes. The European Union (EU) has started to create
the market for such fuels through its mandatory biofuels blending requirements, which will very likely allow electricity-based biofuels to count
towards the blending targets. The implication is that such fuels will no
longer compete against cheap crude-oil-based products, but against biofuels
from palm oil and/or sugar cane, which are significantly more expensive
than crude-oil fuels and less environmentally friendly than electricity-
based liquid fuels.

The potential market opportunity is already material given that only
1% of the total EU liquid-fuel demand of roughly 400-billion litres a year
translates to 4 billion litres a year. In monetary terms, that represents a €3
to €4 billion a year market, based on a biofuels value of €0.8–1.0/l. And
the EU's blending requirement foresees a phased 10% quota for biofuels
over the next years. Conventional biofuels supply roughly 4% into the
EU's liquid-fuel mix today.[23] Even if the EU gradually moves away from
hydrocarbons (diesel and petrol) for ground transportation, its kerosene
market alone, which supplies non-electrifiable aviation demand, is roughly
60 billion litres per year, and growing. There is no other way for Europe
to decarbonise this sector than to make use of power-to-liquids. Given its
advantages, South Africa is arguably in a pole position to be a premier supplier of carbon-neutral synfuels to the EU.

Too hot not to handle

Likewise, there is a genuine opportunity to decarbonise residential, commercial and industrial heating and cooling (while also introducing high levels of electricity system flexibility) by coupling the sector to an electricity
supply industry that is progressively decarbonised and low cost. Yet again,
South Africa has some distinct advantages, mostly owing to the fact that
space heating and cooling, as well as water heating and, in many households,
cooking are largely electrified already. Moreover, many industrial process-
heat solutions have also been converted from coal to electricity, largely
owing to the country's now almost forgotten history of abundant and cheap
electricity (see Introduction). In fact, prior to recent supply disruptions and
surging tariffs, Eskom had an active programme of converting industrial customers from coal to electricity for their heat demand. The load-shedding era
has resulted in some fuel shifting, as households installed solar geysers (i.e.
warm-water boilers) and businesses have sought process-heat alternatives.
Nevertheless, most of the thermal processes in many households (water,
heating, cooling and cooking) remain electrified. Any return to cheap

power would, thus, consolidate electricity's position as the go-to energy source, as was the case in the 1990s. This time round, though, the conversion to electric heating should be pursued not only for its cost advantages, but also for the benefits it can bring in the form of system balancing. In other words, while policymakers should encourage greater energy efficiency, efforts should be made, for the sake of an optimal system, to encourage the electrification of thermal processes in homes, at factories and on farms.

Take residential water heating for example. There is no question that efficiency improvements are possible, and should be encouraged. These could arise in the form of better geyser and piping insulation, or through the installation of heat pumps, which are vastly more efficient at heating water than is a traditional electric geyser. From a system perspective, major benefits could be derived from these electric water-heating installations. This is because, if controlled to dispatch in a system-friendly manner, geysers could become a powerful instrument for sustaining the grid frequency at 50 Hertz, particularly as the share of variable renewable energy rises. Currently, the country's millions-strong geyser fleet is not employed for system balancing. Once the warm water is used, the heating element kicks in to re-heat (i.e. recharge) the geyser. As a result, there is a direct link between warm-water demand and electricity demand. However, it is possible, with today's technologies and without disrupting the supply of warm water, to undo this link and align electric-water-heater loads with the needs of the power system. Decoupling the electric charging from the heat discharging could be achieved either through the creation of an intermediary between households' geyser loads and the power-system operator, or by incrementally installing smart geyser control systems that react to grid frequency.

In the first scenario, the intermediary or intermediaries would gain direct access to and control over, through contracts with homeowners, thousands of geysers. Those loads could then be aggregated and offered to the system operator as an additional grid-balancing tool. Control over only 10 000 geysers would give the aggregator control over a dispatchable load of 30 MW, given that geysers are rated at around 3 kW. Should such aggregation be extended across the country's 6 million geysers, the flexibility contribution could be a whopping 18 GW, which is completely untapped currently – the country's geyser charging comprises about 2 GW of baseload demand, around the clock. In other words, there is significant potential for load shifting, which could increase further should the country's geysers be better insulated to hold the heat for longer. To guarantee consistency of warm water supply, the contract could disallow the aggregator from switching off a geyser when the state of charge is below a certain temperature, such as 55°C to 60°C.

Under the second scenario, South Africa could leverage its geyser fleet through the progressive, but mandatory, integration of intelligent control systems with electric water heaters and heat pumps. These 'plug and play' units would respond to the grid frequency by autonomously charging geysers when frequency rises above 50 Hz and switching them off when the frequency starts to drop. To ensure that homeowners are safeguarded from any negative effects of such switching and that the system does not become unstable because of too many geysers autonomously switching at the same time, the control logic will need to be multi-dimensional. The software would not only need to be attuned to the grid frequency, but would also need to be able to assess whether or not the geyser is on and available to lend grid-frequency support. If so, it should only become available at a point where the temperature of the water is above a certain threshold. Should such an autonomous solution be approved by policymakers and supported by the system operator, there could well be industrialisation spinoffs, as well as export potential. Similar flexibility potential exists in the area of residential space heating and cooling, where solutions are already available to begin breaking the direct link between electricity demand and the thermal service being sought. In addition, the further electrification of industrial processes, on the back of cheap renewables, could offer twin decarbonisation and flexibility benefits.

The conceptual outline of a potential future energy system for South Africa, largely based on variable renewable electricity as a primary energy source, mixed with flexible renewables and flexible natural gas, is outlined in Figure 4.4. In this graph it is assumed that the country would still require 2 350 PJ per year in terms of final end-use energy services, but the losses are significantly reduced to now only 950 PJ, because of the more efficient conversion cycles of a wind- and solar-PV-electricity-based energy system: 3 940 PJ of primary energy (including 250 PJ of ambient heat) would feed this energy system, 65% of which would come from solar PV (180 GW) and wind (120 GW). The CO_2 emissions of this new energy system would be down by more than 95% compared to 2016 to roughly 20 million tonnes per year,[24] while South Africa would export 180 PJ of carbon-free Power-to-Liquid fuels per year (5–6 billion litres worth 50–60 billion Rand per year), 140 PJ of carbon-free fertiliser (3 million tonnes of ammonia worth 20–30 billion Rand per year) and would produce 10 000 bn litres of fresh water per year (the equivalent of today's agricultural demand) from 126 PJ/a of renewable electricity feeding seawater desalination plants. Further, this energy system would be almost entirely independent from imports. Almost all oil imports (today: roughly 18 billion litres worth more than 100 billion Rand per year) are replaced by electric vehicles (battery and hydrogen fuel cell) or by synthetically produced hydrocarbons. The trade balance of South Africa would hence massively improve.

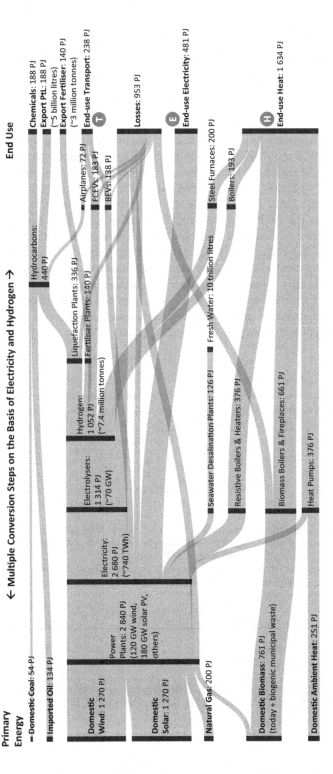

Primary Energy

← Multiple Conversion Steps on the Basis of Electricity and Hydrogen →

End Use

Chemicals: 188 PJ
Export PtL: 188 PJ (~5 billion litres)
Export Fertiliser: 140 PJ (~3 million tonnes)
End-use Transport: 238 PJ
Losses: 953 PJ
End-use Electricity: 481 PJ
End-use Heat: 1 634 PJ

Hydrocarbons: 440 PJ
Airplanes: 72 PJ
FCEVs: 183 PJ
BEVs: 138 PJ
Steel Furnaces: 200 PJ
Boilers: 193 PJ

Liquefaction Plants: 336 PJ
Fertiliser Plants: 140 PJ
Fresh Water: 10 trillion litres

Hydrogen: 1 052 PJ (~7.4 million tonnes)
Seawater Desalination Plants: 126 PJ
Resistive Boilers & Heaters: 376 PJ
Biomass Boilers & Fireplaces: 661 PJ

Electrolysers: 1 314 PJ (~70 GW)
Heat Pumps: 376 PJ

Electricity: 2 680 PJ (~740 TWh)

Domestic Wind: 1 270 PJ
Power Plants: 2 840 PJ (120 GW wind, 180 GW solar PV, others)

Domestic Solar: 1 270 PJ

Natural Gas: 200 PJ

Domestic Coal: 54 PJ
Imported Oil: 134 PJ

Domestic Biomass: 761 PJ (today + biogenic municipal waste)

Domestic Ambient Heat: 251 PJ

Figure 4.4 Potential structure of the South African energy system in the future

As a positive side effect, the throughput of Sasol's re-purposed lique-faction plants would remain almost the same as what it is today at above 300 PJ/a (compared Figures 4.2 and 4.4), i.e. the risk of otherwise stranded assets for Sasol would be fully addressed. Should South Africa fully embrace the scenarios outlined previously, the energy and industrialisation spinoffs could be significant. Under the best-case scenario, the country could repo-sition itself as the Saudi Arabia of green fuels and chemicals such as fer-tiliser; products that could well attract a market premium. However, even if such an aspiration proves commercially challenging, there is still plenty of scope for South Africa to pursue an electrification-of-almost-everything strategy on the back of a low-cost, decarbonised electricity generation. Indeed, access to abundant, clean and cheap power can be the catalyst for a virtuous cycle, whereby the progressive electrification of transportation and heating can yield reductions in carbon emissions and energy service costs, while creating new opportunities for entrepreneurship and industri-alisation. Extracting maximum value, however, will require clear-sighted policymaking, diligent implementation and ongoing, yet focused, research, development and technology commercialisation.

Bottom Line: South Africa can progressively decarbonise its energy sys-tem by coupling its low-cost renewables-led electricity generation to the transportation and heating sectors. At the same time it could create new industrial opportunities and even become the Saudi Arabia of low-carbon, hydrogen-rich chemicals and synthetic fuels.

Notes

1 Brynjolfsson, E. and McAfee, A. *The Second Machine Age: Work, Progress, and Prosperity in a Time of Brilliant Technologies*, New York: W. W. Norton & Com-pany, 2014.
2 International Energy Agency. *World Energy Outlook 2017*, November 2017.
3 International Energy Agency. *World Energy Outlook 2017*, November 2017.
4 United Nations,. *Paris Agreement*, 12 December 2015.
5 LibreTexts Website, accessed October 2017, https://chem.libretexts.org/Core/ Physical_and_Theoretical_Chemistry/Thermodynamics/Thermodynamic_ Cycles/Carnot_Cycle.
6 Using the 'physical energy accounting method' employed by the IEA and oth-ers, as well as the 'direct equivalent method' used by the Intergovernmental Panel on Climate Change.
7 Based on: Comments on the Green Paper 'Ein Strommarkt für die Ener-giewende', Tech. rep., Fraunhofer-Institut für Windenergie und Energiesystem-technik (IWES), status 16 June 2015. www.bmwi.de/BMWi/Redaktion/PDF/ Stellungnahmen-Gruenbuch/150226-fraunhofer-iwes energiesystemtechnik,pr operty=pdf,bereich=bmwi2012,sprache=de,rwb=true.pdf, 2015.

8 Own analysis based on raw data from the IEA. www.iea.org/statistics/statisticssearch/report/?country=SOUTHAFRIC&product=balances&year=2015.

9 *International Energy Agency: Energy Technology Perspectives*, 2017.

10 *South African Petroleum Industry Association Annual Report*, 2014.

11 *CIA World Factbook*. www.cia.gov/library/publications/the-world-factbook/geos/sf.html.

12 Bloomberg New Energy Finance. *Electric Vehicle Outlook 2017*, July 2017.

13 Transaction Capital presentation. SA taxi market, 2016.

14 South African National Taxi Council. www.santaco.co.za/santaco-history/.

15 Transaction Capital presentation. SA taxi market, 2016.

16 International Energy Agency. *Insight Paper: Renewable Energy for Industry*, November 2017.

17 *Agora Energiewende:*. www.cleanenergywire.org/news/synthetic-fuels-keydecarbonisation-study-enbw-invests-taiwan/selective-use-synthetic-fuelsessential-help decarbonise-german-transport-industry-and-heating-thinktanks.

18 *Agora Verkehrswende*. www.agora-verkehrswende.de/en/12-insights/carbonneutralfuels-can-supplement-wind-and-solar-energy/.

19 Prognos on behalf of MWV, Status and Perspectives of Liquid Energy Sources in the Energy Transition, https://www.prognos.com/uploads/tx_atwpubdb/Prognos-summary_phase_I_Liquid_Fuels_as_of_Oct_26_2017-EN.pdf.

20 International Energy Agency. *Insight Paper: Renewable Energy for Industry*, November 2017.

21 International Energy Agency. *Insight Paper: Renewable Energy for Industry*, November 2017.

22 International Energy Agency. *Insight Paper: Renewable Energy for Industry*, November 2017.

23 *Eurostat: Oil and Petroleum Products – a Statistical Overview*. http://ec.europa.eu/eurostat/statistics-explained/index.php/Oil_and_petroleum_products_-_a_statistical_overview.

24 *Emission Factors of Fossil Fuels*. www.umweltbundesamt.de/sites/default/files/medien/1968/publikationen/co2_emission_factors_for_fossil_fuels_correction.pdf.

5 No showstoppers

'What happens to a solar photovoltaic-powered country during a massive solar eclipse? We find out tomorrow. . . ' read one headline on 19 March 2015, the day before the Moon's umbral shadow swept across much of Europe. The darkness would endure for only a couple of minutes for those locations along the path of totality. Nevertheless, it was still the first time in human history that an eclipse was set to have any significant impact on the power system. And system operators were anxious. In Italy, the grid operator decided to mitigate the risk by turning off all large-scale solar plants for the day. In Germany, which at the time boasted more than a quarter of all the solar photovoltaic (PV) electric capacity installed globally, the response was different. The system operator there saw it as an opportunity to learn how to respond to large-scale power variability with speed. Its 1.4-million solar installations, many of them on rooftops, produced around 7% of the country's electricity at the time and lingering questions relating to the risks posed by rising variable renewable-energy penetration persisted.

As it turned out, that Friday's eclipse was indeed a major event. Millions of people across Britain and Europe donned protective eyewear to view the impressive natural phenomenon. And because the 20 March event coincided with the proliferation of social media, the eclipse was shared – through photos and commentary – far more widely than any of the Earth's previous three-billion-odd total eclipses. One topic that failed to trend on the day, however, was anything related to power stability in Germany. The reason, as a *Scientific American* article[1] narrated a few days after the event, was that, following months of planning and weeks of apprehension, there were no blackouts or huge power fluctuations to report. Germany had passed through the eclipse without incident; despite experiencing an almost 9 GW drop in solar input, which fell from 13.5 GW immediately before the eclipse to a low point of 4.8 GW in a manner of less than an hour. One hour after the eclipse began, solar output started to recover again and quickly ramped up by 13 GW. Across Europe, the quick loss of roughly

17 GW was followed by an even speedier reintegration of 25 GW of solar generation, the *Scientific American* article reported.

The key lesson from the 2015 eclipse was not so much about preparing for Europe's next total eclipse, due on 12 August 2026. Rather, it offered something of a template for how system operators might respond to rapid supply changes, as variable generation sources became a larger part of the future generation mix. In fact, it was not really the system operator who dealt with the solar eclipse's effect of switching off and on the German and European solar PV fleet in a matter of a bit more than an hour. It was the electricity market and its participants who did it essentially 'on their own'. The system operators observed closely what happened and would have been ready to jump in with additionally procured reserve capacity for that morning, but they did not need to. The 15-minute slices of the electricity market in Germany and the combined forecasting wisdom of all market participants on the supply and demand side was sufficient to deal with the solar eclipse as if it had been a normal day.

Not so long ago, many electrical engineers were still more or less coming to terms with the reality that a 5% renewables penetration was not in fact the 'physical limit'. This constraint had been drummed into all young engineers, including 50Hertz CEO Boris Schucht, appointed in 2010 to lead the transmission network operator responsible for the grid in northern and eastern Germany. 'Nobody believed that integrating more than 5% variable renewable energy in an industrial country such as Germany was possible', Schucht recalls.

Today, however, he argues that integrating 80% renewables poses no problem, with the 50% penetration level having already been breached in his control area. In a 2016 interview with the European media platform Euroactiv,[2] Schucht described the 2015 eclipse as a very interesting experiment:

> We had been concerned up until then about whether we would be able to deal with it. It demonstrated that we have a lot more flexibility than we first thought and that we will be able to integrate photovoltaics into the system. We are well on track to having a system that can accommodate between 70% to 80% variable renewable energy without the need for more flexibility options. What we already have should be able to meet our needs until 2030 or even 2040.

Unstable grid?

Yet, a persistent concern expressed about increasing the share of variable renewable energy in the generation portfolio relates to the anticipated negative impact these generators will have on grid stability. The worry is that it will be difficult to maintain grid stability and to keep an electrical grid frequency of 50 Hz, the standard frequency used in South Africa and many other parts of the world, once variable renewable energy plants breach a certain penetration threshold. Stability in this context relates to the question of the whether system balance can be restored in the case of a major failure, or contingency event, in the absence of the inertial response provided by conventional power stations, such as coal, hydropower and nuclear. This response arises from the fact that the rotating masses of such power stations, which are synchronously connected to the grid, act to slow down the fall in system's frequency in case of an imbalance between power supply and demand.

What will happen, in other words, in a renewables-led system, if a large electrical generator suddenly disconnects and its electrical power is lost instantaneously? How will grid operators restore balance in the milliseconds and seconds after such an event in light of the fact that wind and solar photovoltaic (PV) plants don't offer the synchronously connected inertia required to slowdown the fall in frequency? These are important questions and have to be tackled comprehensively if South Africa aims to transition to an electrical supply industry undergirded by solar PV and wind generators.

To understand both the risks as well as the possible remedies, it is important to first understand how an electrical system is balanced in real time, as well as what role inertia plays in slowing the rate of change of frequency (referred to by grid operators as the RoCoF) during periods of system imbalance. To do so, it can be useful to visualise the entire electrical system as a single spinning shaft, whose rotation is enabled by both synchronous generators (conventional power stations) and non-synchronous generators (e.g. variable renewables that are connected to the grid via power electronics, or, in South Africa's case, the DC-connected hydro station at Cahora Bassa in Mozambique). This mechanical analogy of the power system's inertia is shown in Figure 5.1.

The speed, or frequency, of shaft rotation is determined by the active power provided by the power generators, balanced against the load arising from factories, offices, mines and homes. To maintain a 50 Hz frequency at all times, the grid operator must balance electricity supply and electricity demand at all times. Inevitably, though, unplanned demand- and supply-side events will arise to upset this balance. And when such events occur, the operator needs to know the rate at which the frequency will fall or rise, or the RoCoF. The RoCoF is determined by how much resistance to change in motion, or inertia, is built into the system. The higher the inertial response, the slower the RoCoF, which effectively buys time for other measures to be taken to restore system balance.

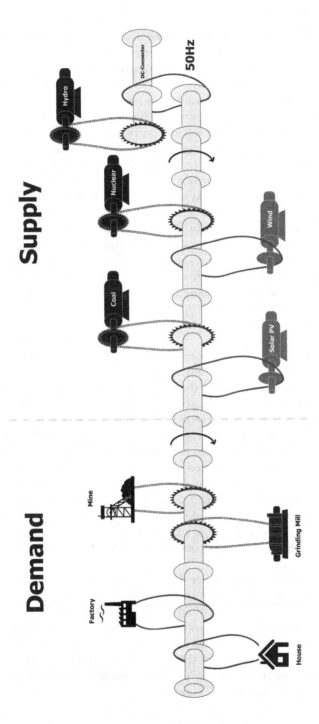

Figure 5.1 Mechanical analogy to demonstrate the effect of 'inertia' in a power system

Returning to our image of the spinning shaft – the constant rotation of which represents stability, or a grid frequency of 50 Hz – system balance is maintained by generators on the one side and the load on the other. Conventional generators are connected to the shaft synchronously by means of chains, while the solar PV and wind generators are connected to the shaft non-synchronously by means of loosely fitting belts. In normal operations, both chains and belts do their job and provide active power to the shaft to keep it spinning. Should one of the generators fail, however, the grid frequency (i.e. the rotational speed of the shaft), will drop immediately, because supply and demand are not in balance anymore. The speed with which the rotational speed drops depends on the remaining synchronously connected generators (the ones connected by chains). Their own kinetic energy will be converted into kinetic energy of the shaft and will reduce the speed at which the rotation slows. What this shows is that synchronous generators inherently provide system stability through the direct, synchronous coupling of their physical inertia to the grid. Put differently, the RoCoF is moderate, providing time for various other grid-balancing counter-measures to kick in.

Timeframes in which a power system reacts to failures are extremely short. After a contingency event, which could arise, for example, in the form of the loss of a large power generator, the grid frequency will start to decrease immediately. Within milliseconds it could have already lost significant ground against the set point of 50 Hz. This in itself is not instantly problematic. The grid frequency has certain limits within which it is allowed to operate without harming general system stability or those end-user devices that require a narrow frequency band in which to operate properly. What is important, though, is for the frequency drop to be slowed down, arrested at the so-called nadir (the frequency at which the change of frequency changes its sign), and finally brought back to the set point of 50 Hz. In most power systems there are three layers of reserves that ensure this in a hand-in-glove type of setting (refer to Figure 5.2). They are not always named precisely the same way across power systems, but the concept is similar across different power systems.

Primary reserves kick in to slow down the RoCoF and to arrest the frequency drop. Primary reserves usually start to supply active power when the grid frequency drops below 49.95 Hz, or when it rises above 50.05 Hz. Primary reserves are usually provided by conventional power generators. Increasingly, though, batteries are playing that role, because of their superior technical characteristics for this particular job. Primary reserves can supply power very quickly, but are usually only required to do so for a limited period of time, of say 5 to 15 minutes.

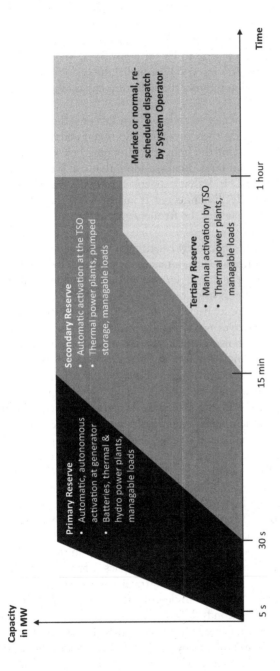

Capacity in MW

Primary Reserve
- Automatic, autonomous activation at generator
- Batteries, thermal & hydro power plants, managable loads

Secondary Reserve
- Automatic activation at the TSO
- Thermal power plants, pumped storage, managable loads

Tertiary Reserve
- Manual activation by TSO
- Thermal power plants, managable loads

Market or normal, re-scheduled dispatch by System Operator

5 s   30 s   15 min   1 hour   **Time**

Figure 5.2 Reserve options in a power system to restore the grid frequency after a contingency event[3]

Note: based on RWE Facts and Figures 2008, page 123, http://rwecom.online-report.eu/factbook/en/servicepages/downloads/files/entire_rwecom_fact08.pdf

During this time, secondary reserves are triggered and take over from primary reserves. Secondary reserves have to be able to keep their power level for longer durations of up to 30 minutes or an hour. They start to move the grid frequency back to its set point of 50 Hz, and then hand over to tertiary or minute reserves. Eventually, the grid is restored and the normal market operation resumes through intra-day trading of supply and demand products.

In the event that primary reserves are not fast or sufficient enough to arrest the grid frequency, then there are additional counter-measures in a power system to prevent the frequency dropping below unrecoverable levels. For example, automatic load shedding that can kick in if the frequency falls below 49.2 Hz. If that happens, certain customer supply areas would be automatically disconnected from the grid, their load would instantaneously fall away and the frequency would stop declining and/or even rise again.

All these mechanisms to securely balance the system at any point in time and under any contingency circumstances are designed to match system-specific characteristics. The RoCoF, for example, is a system-internal property. The more heavy, synchronously connected power generators and customer loads (electric motors) are connected to the grid, the lower the RoCoF and the easier it is for all the reserves mechanisms to manage frequency drops securely.

What would occur, however, in a case where shaft rotation is dependent predominantly on non-synchronous generators, as represented by the generators linked by belts? Although a belt is able to drive the shaft, there is slippage, especially when things change quickly. If only non-synchronous generators are connected to the shaft and one of the belts break, the kinetic energy of the non-synchronous generators is not used directly to slow the deceleration of the shaft, as the slowing shaft simply slips through the remaining belts. In the absence of a counter force to keep the shaft rotating, the rate of deceleration, or in the case of the electricity system the RoCoF, can be very high. If the RoCoF is too high, today's counter-measures might not be able to kick in before the frequency has dropped below an unrecoverable point. Therefore, care needs to be taken when designing a system with low inertia, with clear mechanisms for keeping it stable.

It is important to point out that the term 'inertia' in the context of an electricity system is not the same as the physics definition, which is the tendency of objects to keep moving in a straight line at a constant velocity. In the power context, it is used to describe the amount of kinetic energy, measured in megawatt-seconds (MWs), that is stored in the rotating masses of all synchronously connected power generators. The demand for inertia in an electricity system is driven by two parameters. Firstly, the maximum allowable RoCoF, measured in Hertz per second (Hz/s), and, secondly, the largest assumed system contingency, which is the amount of capacity, measured in megawatts (MW), that can be feasibly lost instantaneously. However, the averaging window period for frequency stabilisation is also important. For grid operators, the RoCoF should not exceed a particular threshold within the pre-defined averaging window, with 500 millisecond (ms) typically used for RoCoF. The following parameters are outlined in South Africa's current grid code:

$$E_{kin.(min)} = P_{cont.} \frac{f_n}{2(RoCoF)} + E_{kin(cont.)}$$

Maximum allowed *RoCoF*: 1 Hz/s within first 500 ms
Largest contingency *Pcont.*: 2 400 MW
Grid frequency f_n: 50 Hz
Kinetic energy lost in contingency event $E_{kin(cont.)}$: 5 000 MWs

Using these parameters, the South African electricity network's inertia demand is 65 000 MWs at any given point to sustain a RoCoF of below 1 Hz/s in the first 500 ms. The largest contingency event in South Africa is 2 400 MW, which is considered to be relatively large by international comparison. This is attributed to the fact that the country's generation system is dominated by large six-pack coal-fired power stations and the largest contingency is based on losing a number of units at one of these power stations instantaneously. By comparison, the largest contingency event in the entire European interconnected power system is approximately 3 000 MW, which is the loss of two large nuclear reactors instantaneously. This number is only slightly more than in the South African case, despite the fact that the European interconnected power system is an order of magnitude larger. This is because the largest contingency is not driven by the size of the power system, but by the size of the largest credible failure. And because Eskom has so many mega stations, the largest credible failure is very large.

As discussed earlier, connecting solar PV and wind to the grid will reduce system inertia. How, then, would a renewables-led generation system meet the inertia requirements of the South African system? Technical solutions are already available to enable the stable operation of systems where inertia is falling, which will occur as renewables plants become a larger part of the generation mix. Some of these solutions have been deployed in island markets, such as the Canary Islands and Ireland, where non-synchronous generation as a proportion of overall generation is relatively high. In the case of Ireland and Northern Ireland, the interconnection to the UK grid cannot provide the inertia required, owing to the fact that the undersea connection is by way of an HVDC interconnector. For this reason, EirGrid and SONI, the transmission system operators in Ireland and Northern Ireland, respectively, have emerged as leaders in dealing with falling inertia, while minimising the costs.[4] These operators are turning to a range of solutions, from intelligent synchronous inertia scheduling formulations to pseudo-inertial control from non-synchronous technology.

However, what would the technical and financial implications be if South Africa simply adopted existing solutions and didn't even consider developments in ancillary services, such as the use of power electronics, to deal with the problem? Using a conservative approach of only deploying additional intrinsic inertia,[5] or synchronous machines, to reduce RoCoF, CSIR researchers have been able to point out a number of technical solutions for increasing inertia in a system dominated by renewables. The list of solutions include the following: synchronous compensators, such as purpose-built devices and retrofitting of decommissioned generators, with or without flywheels; rotating stabiliser devices, such as a multipole device, incorporating a flywheel, that can be based on a doubly-fed induction generator or a synchronous machine; wind turbines with a doubly-fed induction generator; pumped hydro, when synchronous machines are deployed; the 'parking' of conventional generators, which would involve operating a generation plant at low output levels, but with reduced or no capability to provide system services at the lower output levels; reduction in the minimum generation thresholds of conventional generation, while still leaving the plant with the capability to provide system services; and new flexible thermal power plants with a high inertia constant, such as gas-fired power stations, which are significantly heavier, i.e. they have a higher inertia constant, than coal and nuclear.

It is one thing to identify the technical remedies, but what will these cost the electricity consumer? Again, adopting a conservative approach, the CSIR modelled the financial implications using state-of-the-art existing technology, while assuming that there would be no further engineering progress on how to deal with low-inertia systems by 2050.[6] Under this 'worst-case scenario', the additional cost to meet the system's inertia requirements is well below 1% of the total cost of power generation by 2050.[7] In other words, inertia is no showstopper. The technical solutions exist today, and they are not too expensive to destroy the superior economics of renewables. Any engineering advancements until 2050 will make the case even better.

Gas guzzler?

Apart from grid instability, a second concern commonly expressed about a generation mix based on renewable generators is how the system will be 'backed up', particularly during periods when the sun isn't shining, or when wind fails to blow. This is a valid concern, as there are certainly times when neither sun nor wind contribute to the overall supply in a power system. For example during a wind-still night. The first port of call was discussed and described in Chapter 3, where such situations are remedied using flexible, dispatchable power generators that have on-site fuel, such as gas-fired power stations.

However, if gas is proposed even as a bridging solution while alternative storage technologies are commercialised, it is more than justifiable to ask questions about how much gas will be needed and whether South Africa has sufficient gas to plug the gaps. From a financial perspective, too, South Africa needs to understand whether the proposed transition to renewables could translate into a gas-guzzling electricity sector, whereby a dollar-denominated commodity displaces coal, a primary energy source that South Africa has in domestic abundance.

Before answering these questions, it has to be stressed that the cost of primary energy is not the main driver when considering flexibility. Here, the objective is to have generators that can be dispatched at short notice to address any gaps created by the variable production profiles of the weather-dispatched power stations. The profile of a flexible plant is, therefore, quite distinct from that of a conventional baseload power station and its integration into the system is not based on it being the least-cost bulk energy generator. Conventional baseload plants are typically capital intensive, but the high capital costs are spread over many years of high-energy generation, which is produced using an inexpensive primary-energy source. A flexible plant, by contrast, is capital-light, quick and cheap to build and is expected to have a low capacity factor. The variable cost component per energy unit will be high, as will be the flexible plant's total levelised cost of energy when compared with the renewables workhorses in the network. However, the aim is to use these plants only to fill the gaps left by the workhorses, not to *be* the workhorses. Their total energy utilisation, therefore, will be low.

When modelling the gaps that will arise in a South African electricity system where solar PV and onshore wind are the workhorses, the residual load profile can in no way be supplied economically by baseload generators. On some days, the gap between supply and demand will transition from 0 to 20 GW in a matter of hours, and return to 0 just hours later. Even if it was technically possible for this profile to be met through a coal-fired power station, the business case would prove unattractive, as the plant's utilisation would be too low to justify the investment. The gas-fired power plant filling the same gaps would of course also be utilised in the same low manner. But here the difference in cost structure kicks in: gas-fired power stations are capital-light and fuel-intensive. The exact opposite of coal or nuclear. This is a cost structure that is much better suited for low utilisation.

By way of illustration, consider a load profile with a constant demand of 1 GW, and a second load profile where the maximum demand is also 1 GW, but that demand is going up and down. In the first case, it will lead to 24 GWh of energy consumed per day, while the variable demand profile might only draw 6 GWh per day. So, while the installed capacity required is the same for both baseload and flexible demand profiles, in the latter case it will produce far less energy over a day, a month or a year than the former. From an economic perspective, the capital expenditure (capex) for the *capacity* (GW) to supply the load should be low, because the capacity will not be utilised extensively. The fuel costs are less relevant, because the *energy* (GWh) output during a certain time period is small.

The analogy of a trucking company could be useful in explaining the financial implications. If a fleet owner is operating trucks consistently, day and night, the cost of the vehicle becomes far less relevant than fuel costs and the cost of drivers. In other words, fuel consumption should be low even if the vehicle itself is expensive. However, should the truck be used only intermittently, then the fuel costs become far less relevant than the vehicle's initial purchase cost.

Once gas has been selected as the most feasible immediate gap-filler, given that batteries are only likely to prove cost effective later in their commercialisation cycle, the earlier questions of how much gas and at what cost have to be answered. First up, there are no real technical constraints in using gas to provide the flexibility that is envisaged in the renewables-led mix outlined previously. Gas-fired power stations are already playing that role in many countries.

To understand the techno-economics of a generation mix underpinned by solar and wind, but backed up primarily by gas, it may be useful to return to the thought experiment outlined in Chapter 3. There, a theoretical baseload demand of 70 TWh a year was assumed,[8] which was supplied through a generation mix comprising 7 GW of solar PV and 19 GW of onshore wind, backed up by 8 GW of flexible power, arising in the form of natural gas, biogas, pumped hydro, hydropower, concentrated solar power or demand-side interventions. The analysis showed that it was indeed technically possible for such a mix to supply the 8 GW baseload theorised in as reliable a manner as would have been the case using conventional baseload generators. Conservatively, such a mix was estimated to cost 90c/kWh (2016 South African Randcents), which was 10–15% cheaper than the next lowest-cost alternative, being a new-build solution based on coal. The analysis showed that solar PV and wind would supply 89% of total demand, with the 8 GW of flexible power generator making up the rest, and running at an average capacity factor of only 11%. In the event that all of this flexibility was to be derived from natural gas, between 70 petajoules (PJ) and

80 PJ a year of the commodity would be required. This is the equivalent of approximately 1.5 million metric tonnes a year (mmtpa) of imported Liquefied Natural Gas (LNG).[9]

Back to the real world and to a country that does not have readily available domestic gas, the question arises whether it is technically feasible and economically viable to introduce flexibility on the back of imported LNG. Although South Africa is believed to be endowed with shale-gas resources, these resources (estimated to be between 20 trillion cubic feet (tcf) and over 400 tcf) have not yet been proved viable. There are also no firm plans for the construction of a pipeline that would enable South Africa to draw on the world-scale gas reserves that have been discovered in northern Mozambique. Both may offer future opportunity, with South Africa already successfully importing more than 120 PJ yearly through the 865 km Rompco pipeline from southern Mozambique. The gas is currently used by Sasol at its facilities in Sasolburg and Secunda, to produce electricity, fuels and chemicals.

From a market perspective, there are a few obvious hurdles. The LNG market has matured in recent years to offer both diversified and stable sources of supply at increasingly competitive prices, as more and more suppliers have entered the traded seaborne market. The International Energy Agency reports that the volume and diversity of LNG trade flows are increasing rapidly, with the appearance of new exporting and importing countries. Liquefaction capacity is expected to grow by 160 billion cubic metres (bcm) over the period to 2022, led initially by Australia (30 bcm), but with the largest increase in growth then coming from the US (90 bcm).[10] This increase in supply has been associated with a fall in prices, which has resulted in a rapid growth in the number of countries importing LNG, from 15 in 2005 to 39 by 2017. However, South Africa has not yet built import infrastructure for LNG and the authorities have not yet worked out what scale of imports and infrastructure will be necessary to accommodate a generation system based on variable generators.

Here again, there are no showstoppers. The performance of the wind and solar PV plants, as well as the level of demand, will determine gas offtake. The worst-case scenario would be those periods of the year when demand is high, but when it is neither windy nor sunny. Using a blank-page approach to meet South Africa's current demand of around 250 TWh based on a system that depends on solar PV and onshore wind as the workhorses and gas as the flexible generator, gas would contribute 25 TWh a year, translating to yearly LNG imports of around 5 million tonnes. That would be relatively modest when compared to the LNG import facilities that have been built in other countries to meet their power requirements. For instance, Japan is importing 75 million tonnes of LNG yearly across 32 terminals, while

Spain imports 9 million tonnes a year of LNG through seven import terminals. One LNG terminal in China has a capacity of 6.8 million, which would be the equivalent of meeting South Africa's entire demand should its current electricity need be met solely through solar, wind and gas.

Assuming new natural-gas demand of 70 PJ/a (~8 TWh/a of electricity, ~1.5 mmtpa of LNG) to absorb all variability and modelling the hourly gas-demand, the outcome shows that both the scale of LNG storage, and meeting the required state of charge of the LNG facilities, is more than manageable. What would be required is a single 170 000 m³ LNG facility that is refuelled on a 12-day rotation during the worst windless and cloudy periods. Adding an additional storage unit alongside the regasification unit would extend the refuelling schedule to around 20 days. Such an investment would raise the price of electricity produced from LNG only marginally, because the main cost in LNG-power lies in the gas molecules themselves, and not in the gas-storage infrastructure. And even these additional costs would be more than offset by enabling the gas-fired power stations to fulfil their core function of system flexibility. Once seen through a system-wide prism, the cost per kilowatt-hour for a flexible plant becomes far less important, because they are not designed to be the workhorses.

Further modelling by the University of Cape Town to 2050 indicates that the position would still be manageable even in the unlikely event of no cost advances in battery technologies over the period. Under such a scenario, around 47 TWh[11] will have to be produced yearly from gas-fired power stations in a system that is (optimistically) assumed to be almost twice the size, at 435 TWh generated, than it was in 2016. Relatively speaking, therefore, gas-fired power generation would comprise only 11% of the total electricity produced.

However, should battery costs decline and more electric vehicles enter the car parc, as per the least-cost expansion path in Chapter 3, the role of gas would be marginal. While there will still need to be gas-based capacity of 27 GW as a safety net for wind-still, overcast days, the fleet's contribution to overall energy production will be only 11 TWh in 2050, owing to low utilisation rates. Should that energy be supplied purely from imported LNG, only a single import terminal would be required. To put the 11 TWh into perspective, Sasol currently produces 3 TWh of electricity yearly in Sasolburg and Secunda, using gas as an input, and it converts roughly 100 PJ per year into liquid fuels. In fact, 100 PJ is almost the same amount that would be required to produce 11 TWh of electricity per year. Hence, if assuming a gradual move away from the very carbon-intensive gas-to-liquid process, South Africa could simply repurpose the natural gas currently used for electricity production, keeping the volumes constant.

What's more, the model shows that gas may only be a bridging technology for around 20 years as other flexible solutions, such as batteries, become more competitive. Even in the shorter term there are a number of non-natural-gas alternatives to consider, including concentrated solar power (CSP), biogas and pumped hydro. Given that these alternatives are all domestic, policymakers could decide to deviate from the least-cost expansion path to accommodate these solutions. For instance, should CSP be selected to provide flexibility and its costs fall from R2/kWh today to R1/kWh, the technology will be able to compete with batteries and gas. A similar policy choice could be made for biogas, should South Africa perceive there to be a socioeconomic advantage for integrating electricity into the agricultural sector as a revenue-protection mechanism. In other words, total system costs may rise marginally, but the country would have domesticated its fuel supply for the firm capacity.

It's arguably possible for South Africa to even consider converting its baseload contract with the Cahora Bassa hydropower plant, in Mozambique, to one that offers system flexibility. This would mean 'spilling' water rather than feeding it through the turbines. Hidroeléctrica de Cahora Bassa would then be compensated for the provision of firm capacity in Rand per megawatt per day, rather than paying for kilowatt-hours generated. Such a proposal, taken in isolation, appears wasteful in the extreme. However, assessed through the lens of the overall system, it may well be more cost-effective to spill water and accommodate cheaper solar PV and wind energy than to build an alternative flexibility option, such as deploying batteries or gas-fired power stations.

In addition to the flexibility that can be provided by gas and other emerging technologies, there is also significant potential to introduce flexibility on the demand side, as has been discussed in Chapters 3 and 4. Some of the key options include an intelligent dispatch of electric geysers, industrial heat demand and pumping loads. In fact, two good candidates for making electric demand flexible is wherever heat or pumping demand exists. This is because they are typically linked to subsequent processes where the buffer capacity is high. For instance, heating water in a geyser can be uncoupled from the demand for hot water, while pumping water is often associated with water reservoirs that are a natural buffer.

No wind and no sun for prolonged periods?

During prolonged periods of low wind speeds and cloudy weather, a wind-PV-led power system will have to source its energy from somewhere else. The task is in principle no different to the task of providing firm power for the one hour in the evening when there is high electricity demand and no wind. If there is enough firm capacity in the form of gas-fired power stations, or other options, to supply throughout the hour, then there is also enough capacity to supply the customer demand during say two weeks of no wind. What would need to be ensured in this case is not the availability of the firm capacity as such, but sufficient fuel-storage in order to make it through the two-week sunless and wind-still period. As with just about all the scenarios discussed so far, this is an economic, rather than a technical consideration. Gas can be stored in form of LNG in very large quantities and for long periods. This would be the first port of call for such periods. The cost of the LNG storage must be added to the system cost, and during these two weeks the cost to operate the system would be high. But during the other 50 weeks of the year they are extremely low, owing to the low cost of the bulk energy providers, which would be solar PV and wind. What matters from an economic perspective is the long-term annual average, and not the potentially high cost of a few weeks per year.

Technology disruption?

Finally, what about future technological disruption? Is the solar PV, wind and gas mix robust against such a prospect? In a fast-moving world, where there are indeed many technological innovations on the horizon, it is preferable not to be locked in to a specific technology for too long. However, even when considering this dimension, the solar PV, wind and gas mix has material advantages. The key reason for this is that the economic lifetime of such assets is relatively short. The typical business case for solar PV and wind assets is calculated over 20 years, with any energy produced thereafter being 'for free'. Therefore, in year 21, it is a relatively easy decision to make whether to invest in new solar PV and wind plants to replace the old solar PV and wind farms, or to replace them with something new – perhaps even a technology that is not currently known. On the complementary gas-side, the scenario is similar, as such plants also have relatively short lock-in periods. The cost lies mainly with the natural gas itself, and not so much with the infrastructure and the power stations. If an operator stops burning gas, the biggest cost component likewise ceases. The scenario is quite the opposite for coal and/or nuclear power stations. Such plants have much longer economic lifetimes over which to repay the initial capital outlay. Because they are fuel-light it is not economical to shut them down during the capital-repayment period in case demand falls away or superior technologies emerge. Therefore, the wind, solar PV and gas mix is better positioned to adjust to any future technological disruption.

Bottom Line: The proposed energy transition for South Africa is an enormous task. However, the good news is that there are no technical showstoppers to the transition. Any and all foreseeable technical challenges can be addressed with today's available technologies and procedures, and at costs that are negligible compared to the total cost of a power system. Any progress in engineering solutions for flexible and renewables-based power and energy systems will make the outcomes even better.

Notes

1 Marx, E. Scientific America Website, accessed October 2017, https://www.scientificamerican.com/article/how-solar-heavy-europe-avoided-a-blackout-during-total-eclipse/.

2 Dehmer, D (translated by Morgan Sam) on EURACTIV Website, accessed September 2017, https://www.euractiv.com/section/energy/interview/german-electricity-transmission-ceo-80-renewables-is-no-problem/.

3 http://rwecom.online-report.eu/factbook/en/pics/img/4_gen_bal_power_en.png

4 Daly, P. *Presentation on Power System Inertia: Challenges and Solutions*, 2016.

5 Bischof-Niemz, Dr T. *Energy Modelling for South Africa, Latest Approaches & Results in a Rapidly Changing Energy Environment, Presentation to the INCOSE Western Cape Branch*, 2017.

6 Chown, Graeme, Wright, Jarrad, van Heerden, Robbie and Coker, Mike. *System Inertia and Rate of Change of Frequency (RoCoF) with Increasing Non-Synchronous Renewable Energy Penetration*, 2017.

7 Bischof-Niemz, Dr T. *Energy Modelling for South Africa, Latest Approaches & Results in a Rapidly Changing Energy Environment, Presentation to the INCOSE Western Cape Branch*, 2017.

8 Bischof-Niemz, T. *The Case for Renewable Energy to Provide Base Load Energy in South Africa*. Presentation to Presentation at the Sustainability Week 2017, June 2017.

9 Bischof-Niemz, T. *What Is the Role for Gas and Renewables in RSA?* Presentation at the Gas Africa 2017 conference, May 2017.

10 International Energy Agency. *Market Report Series: Gas 2017*, 2017.

11 Gregory Ireland and Jesse Burton, University of Cape Town, Energy Research Centre, An assessment of new coal plants in South Africa's electricity future, page 23, http://www.ee.co.za/wp-content/uploads/2018/05/ERC-Coal-IPP-Study-Report-Finalv2-290518.pdf.

6 Democratisation of electricity generation

With the lights dimmed in a crowded room at a conference centre in northern Johannesburg, two large screens offer dazzling illumination amid the gloom. The atmosphere is heavy and the presenter is focusing almost all of his delivery attention towards the four individuals sitting behind a royal blue cloth-draped table, branded with the energy regulator's 'Nersa' logo. Neatly packaged within the mandatory blue and white template used for all Eskom PowerPoint presentations, the slide being projected is one of a graph depicting the utility's electricity sales since 1994. The illustration includes three distinct lines, one red, one green and one black. The red line represents actual sales over the period, the green the assumed sales forecast for the five-year period from 2013 to 2017, while the black line offers a new sales forecast to 2027. From 1994 to 2007, the red line rises steeply from below 150 TWh/a to a peak of above 224 TWh/a. It then retreats sharply, before recovering meekly. The red 'actual sales' line then falls gently to join up with the black line, labelled 'realistic forecast'. The black line tracks sideways, never again breaching the 220 TWh/a level. The green line, which represents the previously approved forecast, offers a stark illustration of just how far expectations have strayed from reality. Its steep trajectory, rising to over 244 TWh/a by 2017, appears both naïve and unrealistic – a graphic reminder to all the gloomy onlookers in the room, of just how wrong planners and regulators were in 2012.

The slide is also probably the clearest sign that Eskom's business model is being disrupted, even in the absence of any major policy shift. On the supply side, this disruption is being driven by the falling costs of generation technologies that can be introduced even by an individual homeowner. On the demand side, energy efficiency, together with a generalised decoupling of economic growth from energy use, is having profound effects on the utility model. While demand falls, the same is not true for costs, which continue to rise, creating a vicious cycle that many term the 'utility death spiral'. In the South African context, there is particular concern that the price elasticity of demand has become an important factor in dampening industrial offtake. In other words, the tipping point has either been passed, or is close to being reached, whereby a small rise in electricity tariffs will be accompanied by a steep fall in demand.

Such decoupling of sales and costs is not unique to Eskom and has caused, or is causing, turmoil for incumbent utilities the world over. The most dramatic response has been in Germany, where two large utilities, E.ON and RWE, have already split in an effort to navigate difficulties faced. These included falling demand, the rapid growth in wind and solar capacity and Germany's decision from 2000, and re-confirmed in the wake of Japan's Fukushima disaster in 2011, to prematurely decommission all nuclear power stations by 2022. On 1 January 2016, E.ON separated its conventional power generation and energy trading operations into a new company called Uniper, while retaining its retail, distribution and nuclear operations. In an interview with the *Financial Times* at the time of the split, E.ON CEO Johannes Teyssen declared the integrated model 'dead' and admitted that the company had misjudged the radical speed and nature of the change in the electricity industry. On 1 April 2016, RWE began operating a new subsidiary for renewables, grids and retail, separated from its nuclear, gas and coal business.

A report by advisory group EY, meanwhile, estimates that utilities across Europe wrote off €120-billion of assets because of low power prices between 2010 and 2015.[1] The National Grid's 2017 Future Energy Scenarios, meanwhile, outlines increasingly complex electricity flows, as the dominance of large thermal plants eases and decentralised generation, or generation located on the distribution network rather than the transmission network, increases. The report argues that the trend will continue and that the shift towards decentralised and renewable generation is evident in all its scenarios, with only the pace and extent of this change differing across the scenarios. As a result, the requirement for system flexibility will increase, as the amount of variable and decentralised generation grows.[2]

Yet, this requirement for flexibility does not arise all at once. Analysis conducted in markets at the forefront of integrating variable renewable energy (VRE) into their networks points to four distinct phases. The International Energy Agency[3] says that, during phase one, where VRE has a less than 5% share of generation, the system operator and other power system actors do not need to be concerned with the variability associated with wind and solar photovoltaic (PV). During phase two, of between 5% and 10% VRE penetration, the existing generation fleet will see changes in their generating pattern, but the system can accommodate the new situation largely with existing system resources and/or by upgrading certain operational practices. In phase three, above 10%, system flexibility becomes key for integrating VRE. This flexibility relates to how quickly the power system can respond to changes in the demand and supply balance in a timescale of minutes to several hours. During phase 4, or high VRE penetration rates of between 20% and 50%, the main challenge relates to stability. During this phase, the IEA says improved operational practices are required to maintain power system reliability.

Thought is now already being given to the implications of even higher penetration rates, but market experts stress that the transition between phases will not be abrupt. Instead, issues relating to flexibility will gradually emerge in phase two, before becoming the hallmark of phase three. In turn, certain issues related to system stability may already become apparent in phase three.[4] To operate in the fourth phase, though, requires an entirely different approach from the classic generation dispatch model of baseload first, followed by mid-merit and peaking plants. In fact, Moeller Poeller Engineering cofounder Dr Markus Poeller argues that an entirely new mindset is required, based on variability and flexibility.[5] In this world, baseload as a concept all but disappears and both flexibility and stability factors need to be taken into account by system operators. In such a context, battery storage is likely to play an increasingly important role – for short-term stability purposes more than for the purpose of shifting large amounts of energy from time A to time B.

This fundamental shift in the provision of electricity services will be coupled with consumption changes. The Massachusetts Institute of Technology Energy Initiative's Utility of the Future study says these changes are being underpinned by a confluence of factors, including distributed generation and energy storage, with increasingly powerful information and communications technologies and control systems.[6] To achieve a level playing field between centralised and distributed resources, the study proposes a new framework that includes a comprehensive system of market-determined prices and regulated charges for electricity services that reflect the incremental cost of providing these services. It also suggests a rethinking of the industry structure to minimise conflicts of interest, as well as a redesign of regulation to cater for the new reality.

The technical energy system of the future will be designed around the two new variable workhorses: solar PV and wind. Similarly the regulatory framework will have to be designed around them, too. Just like today's regulatory framework was designed around the inflexibility of today's workhorses coal and nuclear.

Although South Africa is not at the forefront of these changes, it is certainly not immune to them, with the signs of utility stress becoming increasingly visible, amid softening demand and rising electricity tariffs.

Unless these are proactively managed, there could be a restructuring by default, with some major casualties along the way.

Under pressure

Undeniably, the current model is not working for electricity consumers, the economy and even for Eskom. Over the past ten years, the country has been subjected to record tariff hikes, yet these have proved insufficient to fund Eskom's new-build programme and sustain its credit rating. At the same time, poor operational performance has seen Eskom resort, for prolonged periods, to expensive supply options to close a supply gap that would not have arisen at all if the coal-fired fleet's energy availability factor had been sustained at 80% or more.

As if not toxic enough, the crisis has been deepened further by allegations of financial misappropriation and corruption. These factors, together with Eskom's inability to manage an enormous and inflexible capital programme, have entrenched a vicious cycle of cost increases, higher debt (set to peak soon at R500 billion from today's R350 billion) and surging debt repayments, which have increased to around R40 billion a year and will peak at R60 billion in 2021/22. The outcome is yet more hikes, triggering even lower demand, weaker Eskom financials and, ultimately, more government support, in the form of guarantees or even financial bail-outs. Indeed, the utility is increasingly leaning on the taxpayer for funding, which has come in the form of direct capital injections, the conversion of a subordinated loan to equity and guarantees to enable Eskom to raise domestic and foreign debt, notwithstanding its junk rating.

The utility's response to falling renewable energy costs, meanwhile, has been defensive. Instead of assessing what this new reality means for its own capacity expansion plans and overall system costs, Eskom has moved to suppress the entry of new solar PV and onshore wind projects. This response has highlighted the systemic dangers of sustaining a vertically integrated monopoly in a context of rapid and fundamental change. In the final analysis, South Africa has an electricity supply industry model that is at once unsustainable and too big to fail – the worst of all worlds.

Fit for purpose?

It should be stressed, however, that this has not always been the case. South Africa's centralised, state-owned model was more than appropriate for the old electricity world where 'bigger was better' and where carbon emissions were not a constraint. In such a world, it made perfect sense for large coal deposits to be mined by large companies and supplied to a large utility that operated large power stations. The utility burned the coal efficiently to produce bulk electricity, which was transported effectively along large, high-voltage transmission lines, primarily to large industrial and municipal customers. With some exceptions, the electrons flowed in a single direction, from power stations clustered in the northeast, to load centres further south. In such a centralised world, the vertically integrated utility model was not only optimal, but also sustainable and efficient. Indeed, it formed the very basis for South Africa's industrialisation and modernisation, built primarily on the exploitation of the country's formidable hard mineral resources.

In the unfolding 'new world' of both centralised and decentralised generation, and a demand trajectory that is less certain, the model is no longer fit for purpose. In this world, the risks associated with integrating lumpy mega-scale electricity generators are high, as has already been amply demonstrated with the ill-fated Medupi and Kusile projects. What's more, the falling costs of solar PV and onshore wind will mean that the generation fleet will consist of individual power stations that are orders of magnitudes smaller than today's coal plants, but many more of them, and it will be geographically spread. This makes it possible to introduce real competition for new generation investments. Something very difficult in a world made of Medupis and Kusiles.

In such a context, it would be better to have the generators, both big and small, owned and operated separately from the infrastructure that enables market participants to interact with each another. This enabling infrastructure is the grid, which comprises both transmission and distribution networks. Like the motorways and roads in the transport context, the enabling infrastructure should be regulated and can even be state-owned. Using this analogy, it would make little sense for the cars and trucks to be owned and operated by the agency or public institution that oversees the construction and maintenance of the roads.

What does this mean for the South African electricity supply industry? Firstly, it will require the vertical separation of Eskom's generation assets from its network and system operation assets. Today's structure, where the state-owned utility owns and dominates the entire electricity value chain (Figure 6.1), is simply not fit for purpose anymore.

To achieve this, Eskom's generation assets should be sold to new owners (public and/or private entities) through a competitive process. Up for sale would be the asset, together with a power purchase agreement (PPA), aligned with the residual lifetime of the power station. A condition of the PPA should be that money be ring-fenced and set aside for

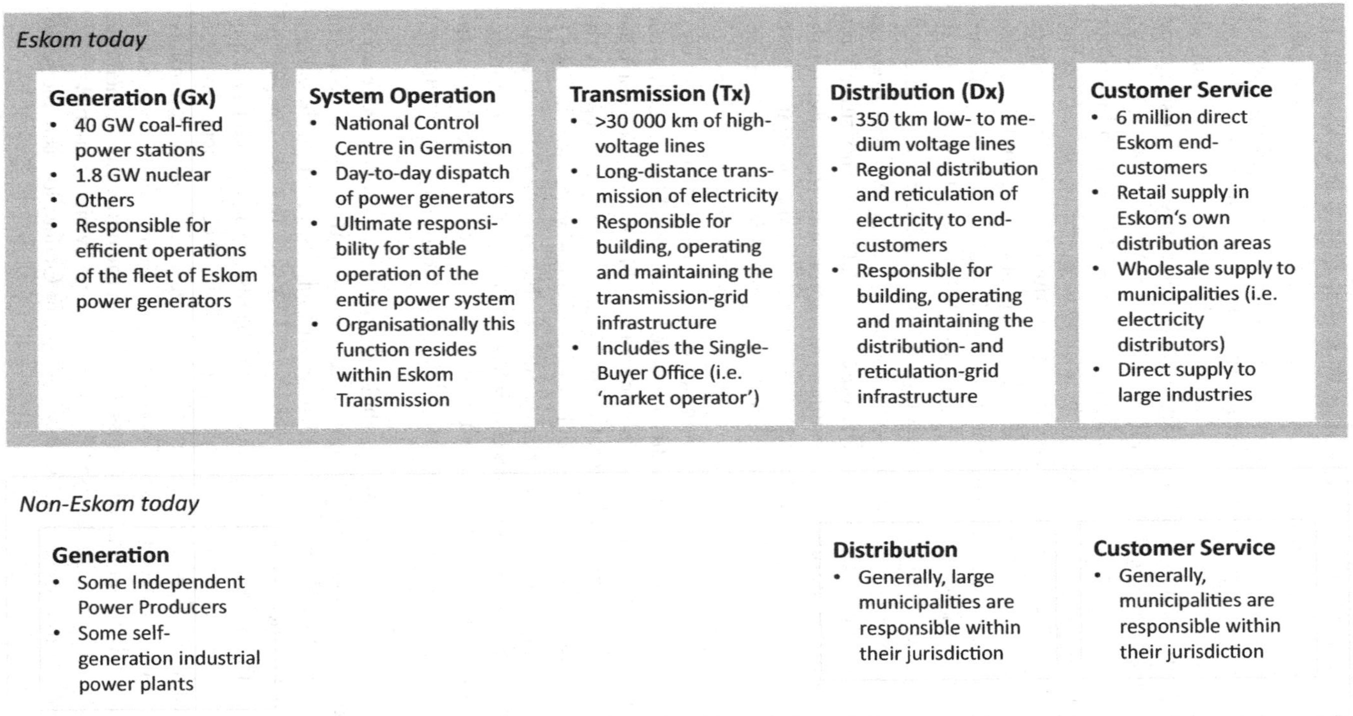

Eskom today

Generation (Gx)
- 40 GW coal-fired power stations
- 1.8 GW nuclear
- Others
- Responsible for efficient operations of the fleet of Eskom power generators

System Operation
- National Control Centre in Germiston
- Day-to-day dispatch of power generators
- Ultimate responsibility for stable operation of the entire power system
- Organisationally this function resides within Eskom Transmission

Transmission (Tx)
- >30 000 km of high-voltage lines
- Long-distance transmission of electricity
- Responsible for building, operating and maintaining the transmission-grid infrastructure
- Includes the Single-Buyer Office (i.e. 'market operator')

Distribution (Dx)
- 350 tkm low- to medium voltage lines
- Regional distribution and reticulation of electricity to end-customers
- Responsible for building, operating and maintaining the distribution- and reticulation-grid infrastructure

Customer Service
- 6 million direct Eskom end-customers
- Retail supply in Eskom's own distribution areas
- Wholesale supply to municipalities (i.e. electricity distributors)
- Direct supply to large industries

Non-Eskom today

Generation
- Some Independent Power Producers
- Some self-generation industrial power plants

Distribution
- Generally, large municipalities are responsible within their jurisdiction

Customer Service
- Generally, municipalities are responsible within their jurisdiction

Figure 6.1 Today's industry structure of South Africa's electricity system

decommissioning and rehabilitation. The entity overseeing the privatisation process should ensure that no individual market player is allowed to own more than 20% of the assets in terms of energy share, as well as in terms of firm capacity, in order to prevent anti-competitive behaviour and a single dominant player. In addition, the tariffs of the PPAs should be predetermined and should comprise two elements. The first component should cover energy, or kilowatt-hours (kWh), while the second should be for capacity, or guaranteed kilowatts. Bidders could then outline how much money they would be willing to invest to buy the package of assets and the associated PPAs, where the duration and tariff is predetermined per power station. All existing coal contracts would be covered by the PPAs, so as to ensure there are no immediate losers as a result of the sales process. Each generation asset should be allocated a certain residual amount of terawatt-hours that it is entitled to produce and sell. This represents both residual coal volumes and total revenues from the energy that will be produced. The outcome would be competitively run coal-fired power stations and an inflow of capital into the coffers of the National Treasury. This inflow of capital would be the present value of the individual PPAs linked to Eskom's coal-generation fleet. Depending on where the pre-defined PPA tariff per power station is pegged, hundreds of billions of Rand could be injected into the fiscus, which could be used to repay Eskom's loans. In addition, government could ensure that the deals are structured in a way that encourages black ownership of what would be secure deals, where future revenues are certain, albeit in the context of assets that will eventually close. In other words, for the sake of business sustainability, the owners would be able to invest their low-risk returns from a gradually ramped-down coal fleet into more sustainable future assets. Likewise, the salary bill and staff numbers would be guaranteed for a period and their present value can be factored into the upfront lump-sum payment.[7]

In the South African context, wholesale privatisation of the generation assets could prove politically unfeasible. However, there is still scope to achieve the desired restructuring with a far higher level of public ownership. For instance, state pension funds could become the new owners and outsource the technical management of the plants, or there could be a stipulation that the state retains 49% of all the individual coal-generation assets separated from Eskom. Yet, the price for the asset should still be determined through a competitive process. The alternative of doing nothing would be an uncontrolled, unmanaged process where both the mining and generation assets, together with Eskom itself, become increasingly unsustainable.

The structural result of such a sale of Eskom Generation is depicted in Figure 6.2.

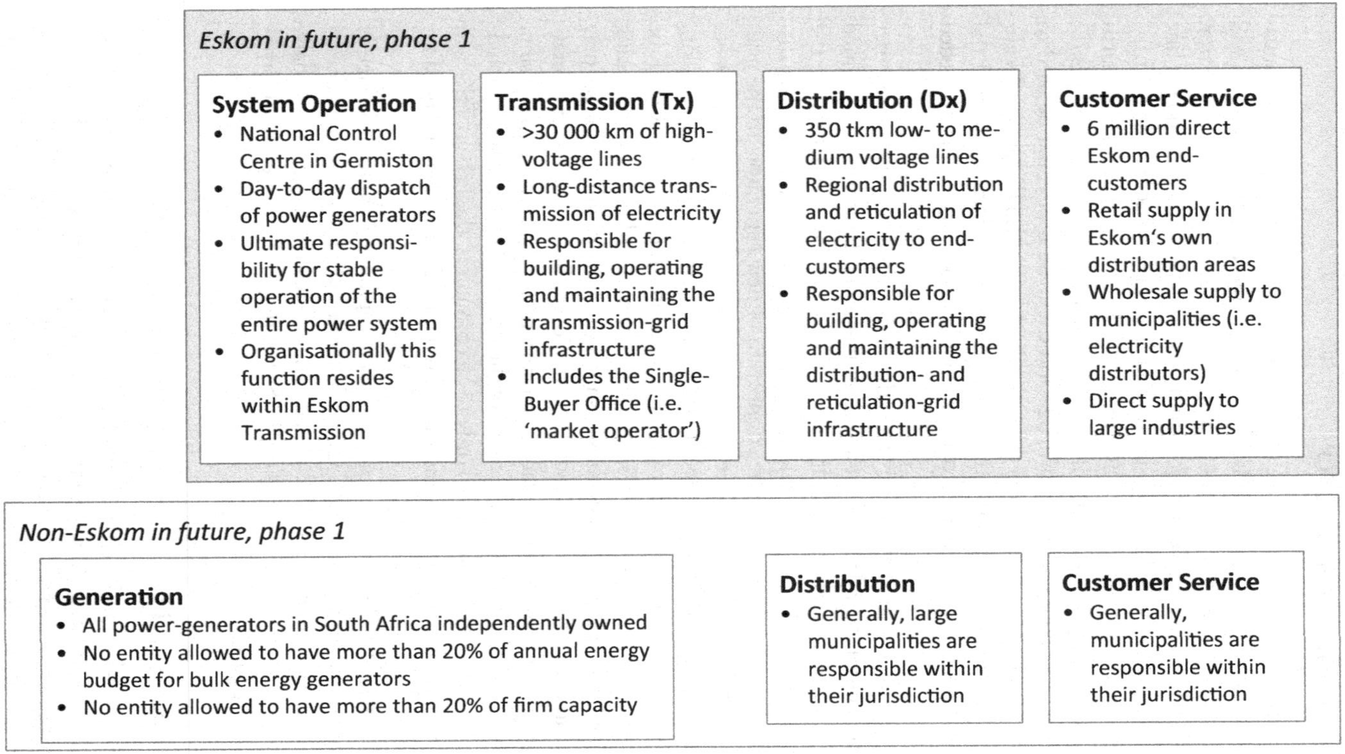

Eskom in future, phase 1

System Operation
- National Control Centre in Germiston
- Day-to-day dispatch of power generators
- Ultimate responsibility for stable operation of the entire power system
- Organisationally this function resides within Eskom Transmission

Transmission (Tx)
- >30 000 km of high-voltage lines
- Long-distance transmission of electricity
- Responsible for building, operating and maintaining the transmission-grid infrastructure
- Includes the Single-Buyer Office (i.e. 'market operator')

Distribution (Dx)
- 350 tkm low- to medium voltage lines
- Regional distribution and reticulation of electricity to end-customers
- Responsible for building, operating and maintaining the distribution- and reticulation-grid infrastructure

Customer Service
- 6 million direct Eskom end-customers
- Retail supply in Eskom's own distribution areas
- Wholesale supply to municipalities (i.e. electricity distributors)
- Direct supply to large industries

Non-Eskom in future, phase 1

Generation
- All power-generators in South Africa independently owned
- No entity allowed to have more than 20% of annual energy budget for bulk energy generators
- No entity allowed to have more than 20% of firm capacity

Distribution
- Generally, large municipalities are responsible within their jurisdiction

Customer Service
- Generally, municipalities are responsible within their jurisdiction

Figure 6.2 Electricity industry structure in South Africa after a sale of Eskom Generation

Grid company: 'Eskom TSO'

Once the generation assets are sold, Eskom will be left with the transmission assets and will be the country's systems operator. In other words, it will become what is known internationally as a transmission system operator, or TSO. It will be responsible for building and maintaining the power line infrastructure required for transporting electrical power across the country. There could even be scope for Eskom TSO to play a regional role in Southern Africa and transform itself into a pan-African super grid company. To be sure, Eskom has the knowledge and expertise that could be used to accelerate electrification across the regions. There is also the greater likelihood that, once a strong infrastructure company is in place throughout the region, private generators will follow Eskom TSO and deploy much-needed generation assets in the rest of Southern Africa.

Crucially, Eskom TSO will oversee the entire electricity system, balancing generation and demand on a second-by-second basis within a defined set of frequency and voltage parameters. The company will determine the optimal combination of generating assets and how these should be dispatched. It will also oversee the connection of new generators to the grid. Eskom TSO is likely to play a central role in planning for long-term system security and provide critical input to the country's integrated resource plan. The single-most important performance indicator for Eskom TSO will be the overall system reliability. This can be achieved without Eskom seeking to secure its own generation market shares, because it will no longer play a role in generation. The Eskom TSO would be the conductor or referee, not the player. As more renewables generation is integrated and the system becomes more democratised in terms of the number of participants, including those previously considered only to be consumers, the Eskom TSO will also have to develop the systems and tools required to address the need for flexibility and stability.

This separation of duties is not 'predator-capitalism' thing. It is merely good and efficient governance. China, interestingly, has separated generation from transmission in 2002 already and so created the largest, globally active grid-infrastructure company / TSO in the world: State Grid Corporation of China.

Distribution grid

At the distribution level, meanwhile, the ownership of the physical assets should probably remain with either Eskom or the municipalities. However, policymakers will need to review the current structure, owing to serious sustainability concerns, particularly among small municipal distributors. In other words, there may be a need to rationalise and consolidate distributors into fewer, more robust, entities, but without infringing on a municipality's right to distribute power.

Distribution grid owners should be empowered to decide whether or not to insource or outsource the upgrading, operation and maintenance of the network. A hybrid model is possible, whereby larger distributors, such as the metropolitan utilities and Eskom, retain ownership and control whereas smaller municipalities could outsource ownership and/or management to a concessionaire (could be larger municipalities or Eskom). Should an outsourced model be selected, the municipality could earn a concession fee simply for making the land and servitudes available. In turn, the asset manager would earn a regulated return, which should be performance linked. Through national benchmarking, poor performers could be weeded out and a rival asset manager introduced.

Retail competition

Eventually, after the successful introduction of competition on the generation side of the electricity supply chain, the introduction of competition on the retail side could be considered, too. Today in South Africa, the physical location of an end-customer determines who the electricity retailer to that particular customer is (either the local municipality or Eskom). With retail competition, electricity customers in South Africa would have a choice of who supplies them with electricity through the monopolistically structured and fully regulated transmission, distribution and reticulation wires. The actual electrons passing through the assets would then no longer be sold by either Eskom or the municipalities, which simply provide the infrastructure. Retail competition means that electrons delivered to end customers can be sold by any number of licensed retailers. These retailers buy the electricity from the wholesale market, which would be a blend of Eskom's now split off coal fleet and independent power producers, complemented with whatever suits their portfolio, such as rooftop PV and batteries.

The retailers would not have to own any physical assets (neither generators nor wires), but they would simply be an efficient trading function between the generators on the side of the value chain and the end customers on the other side. This way on-site, behind-the-meter generators like rooftop solar PV could also be integrated into the entire market structure. In such a structure one would need a market operator that determines the rules of the game. Such a potential future structure is depicted in Figure 6.3.

Without doubt, there would be risks in implementing such a radical restructuring in one go. However, these risks could be mitigated through a

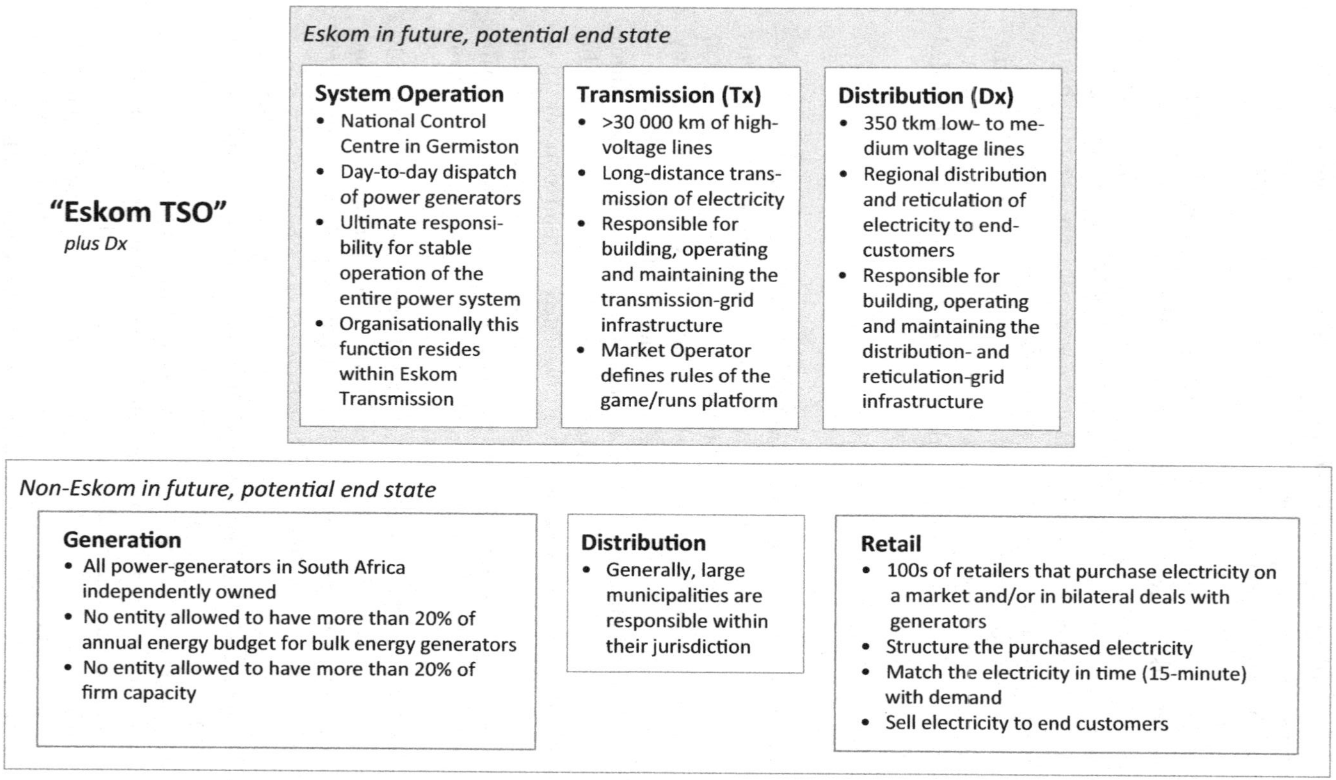

Eskom in future, potential end state

"Eskom TSO"
plus Dx

System Operation
- National Control Centre in Germiston
- Day-to-day dispatch of power generators
- Ultimate responsibility for stable operation of the entire power system
- Organisationally this function resides within Eskom Transmission

Transmission (Tx)
- >30 000 km of high-voltage lines
- Long-distance transmission of electricity
- Responsible for building, operating and maintaining the transmission-grid infrastructure
- Market Operator defines rules of the game/runs platform

Distribution (Dx)
- 350 tkm low- to medium voltage lines
- Regional distribution and reticulation of electricity to end-customers
- Responsible for building, operating and maintaining the distribution- and reticulation-grid infrastructure

Non-Eskom in future, potential end state

Generation
- All power-generators in South Africa independently owned
- No entity allowed to have more than 20% of annual energy budget for bulk energy generators
- No entity allowed to have more than 20% of firm capacity

Distribution
- Generally, large municipalities are responsible within their jurisdiction

Retail
- 100s of retailers that purchase electricity on a market and/or in bilateral deals with generators
- Structure the purchased electricity
- Match the electricity in time (15-minute) with demand
- Sell electricity to end customers

Figure 6.3 Potential end-state of the electricity system in South Africa

managed process, which begins with the separation of Eskom's generation assets from its transmission and system operator units. Once complete, a phased restructuring of the distribution sector could follow, with the objective of improving service delivery and lowering costs through the introduction of retail competition.

Restructuring of the electricity supply industry is not an end in itself. The ultimate goal is to ensure that the energy services delivered to end users in South Africa are both affordable and reliable. Achieving these two objectives within the current framework appears less and less likely. Therefore, a new structure will have to be conceived: a structure that is not only more sustainable, but also more resilient to disruption.

Bottom Line: South Africa's vertically integrated utility model is already showing signs of severe distress. This is occurring while the world moves towards a more distributed energy-supply model. A new model, which separates Eskom's generation assets from its transmission and system operation units, is likely to prove more resilient to the disruption already under way, as well as to any future shocks to the functioning of the electricity supply industry. The government will keep full control of the electricity sector by way of owning and controlling the physical infrastructure that enables the flow of electrons in the first place: the grid. The electrons themselves would be optimally generated by a large number of competing entities, which would drive down costs and increase democratisation of the ownership base of electricity generation.

Notes

1 EY, *Benchmarking European Power and Utility Asset Impairments: Living with Lower for Longer*, 2016, page 4.
2 National Grid, *Future Energy Scenarios*, July 2017, page 55.
3 OECD/IEA, *Status of Power System Transformation 2017: System Integration and Local Grids*, June 2017.
4 OECD/IEA, *Status of Power System Transformation 2017: System Integration and Local Grids*, June 2017.
5 Dr. Markus Poeller, Presentation 'Power Systems with Very Large Wind and Solar Generation – What Is Needed for a Stable and Secure Operation?' at the SA Energy Storage Conference, November 2017.
6 Massachusetts Institute of Technology, *Utility of the Future: An MIT Energy Initiative Response to an Industry in Transition*, December 2016.
7 *Igniting Eskom Generation: Turning the Deadweight into Economic Fuel.* www.ee.co.za/article/igniting-eskom-generation-turning-deadweight-economic-fuel.html.

7 Just transition

In the preface to this book we reflected, through the eyes of a fictional president, who we named Khanyisile Moya, on South Africa's 2050 energy system. Built on the platforms of sound techno-economic principles and visionary policymaking, South Africa had, over a period of three decades, succeeded in decarbonising, de-risking and democratising its electricity supply industry. The country had reached a point where it was not only producing affordable and clean power, but was also using its world-class solar and wind resources as the basis for the manufacture of a range of high-value exports.

Moya attributed this success to the clear-sighted and consistent energy policy decisions of her predecessors. However, she also reflected on the fact that their choices had been highly contested at the time, owing to powerful vested interests embedded in the country's large-scale, coal-based and centralised electricity system. Fortunately for South Africa, its energy leaders had been alive to the dramatic shift in technology costs and what those changes meant for a country so richly endowed with solar and wind resources. In seizing the moment, those leaders had helped South Africa reclaim its position as one of the lowest-cost electricity producers globally. The economic and employment effects were dramatic and enduring. South Africa's rediscovered electricity competitiveness became the cornerstone for a vibrant and employment-intensive industrial policy, centred on the electrification-of-almost-everything vision. By 2050, South Africa had emerged as Africa's manufacturing hub for solar and wind components, as well as a large-scale exporter of electric vehicles. It also produced low-carbon fuels, chemicals and fertilisers, as well as a plethora of energy-intensive metal and engineering products.

But for the vision outlined in Moya's reflections to have even the slimmest fighting chance in reality, serious arguments about jobs and economic transformation have to be made and won. This is especially true in a context where the prevailing narrative around renewable energy has become one of jobs losses and coal-mining ghost towns.

Jobs rich

Indeed, opponents to renewable energy typically question whether the sector can ever realistically create the same number of jobs as a system based on large-scale nuclear or coal. This image of relative job poverty is easy to conjure when individual solar and wind farms are compared with a megaproject. Take the Medupi coal-fired power station, which is about an hour's drive from the 66 MW Tom Burke solar farm. Much like the 4 800 MW coal plant's visual impact, as it rises from what was once undeveloped bushveld, its employment impacts are obvious for all to see. The project has fundamentally altered the Lephalale's socioeconomic landscape. New housing, access roads and other bulk infrastructure have been developed to accommodate the tens of thousands of construction workers entering the site daily, often causing traffic jams. What's more, thousands of permanent jobs are being created at both the power station and Exxaro's nearby Grootegeluk coal mine, as Medupi transitions from building site to power plant. However, it must be stressed that comparing this high-profile impact with that of Tom Burke, which created a few hundred construction jobs, but only a handful of permanent ones, is simply not an apples for apples comparison. That is because Medupi, once fully operational, will generate 250 times more electricity per year than the Tom Burke solar farm (34,000 GWh/a vs. 130 GWh/a). Hence, in this specific case, the job impact of 250 'Tom Burkes' would have to be compared with that of one Medupi.

For a generic comparison, an analysis is required of the job-years involved in installing and operating the different generation technologies, relative to the size and electricity output of the respective plant. And here, research conducted for the Department of Energy by McKinsey & Company for the Integrated Energy Plan (IEP)[1] is instructive. Before interrogating those numbers, it is important to recall that 1 GW of coal, or nuclear capacity, produces far more energy than 1 GW of wind or solar PV. Therefore, much more wind and solar PV capacity is required to produce the same amount of energy. In other words, direct calculations using the raw data will not show the true picture.

The first step, nevertheless, is to understand the raw jobs data. Helpfully these are provided in the IEP. In fact, the McKinsey analysis provides estimates of the direct, supplier, indirect and induced jobs arising from the building and operation of various generation technologies. We will focus on the direct and supplier jobs only, as they are the clearest, least diluted numbers directly linked to a certain economic activity. The technologies include solar PV, concentrated solar power (CSP), onshore wind, nuclear, pulverised coal-fired and associated coal mining, as well as combined cycle gas turbines (CCGT), using shale gas as an input. The analysis assumes that 5 GW of any technology is required in order to ensure an appropriate like-for-like comparison and assesses only new-build job creation and not the effects of changing the technology of existing energy capacity. In addition, assumptions are made regarding the potential to create jobs by locally producing different electricity generation technologies.

Two criteria are used in the IEP to assess this localisation potential, including a sufficiency of demand for a good or service and the ability of the country to potentially supply this particular spend component, rather than import it. Put another way, there is a scale dimension (demand sufficiency) and a time dimension (how long will it take to localise on the back of sufficient demand). For South Africa to fully realise the jobs and industrialisation potential of its electricity build programmes, it is important to understand which technologies have the most localisation potential. The IEP interrogates the value chains of the various generation technologies and quantifies how many supplier jobs are required for each technology.[2] The assessment outlines those inputs associated with each technology that could be immediately produced within South Africa, as well as the localisation potential for those inputs not immediately available in the country. In addition, a scale was applied showing the degree of difficulty involved in localising the various components, over a short or medium term of three to seven years and in the long term of more than seven years. The scale ranges from immediately 'localisable' at 5 GW, or easy to produce locally at that scale, to categories where both significant investment and/or global demand would be required in order to localise.

Interestingly, for both solar PV and wind, the immediate localisation potential was relatively high at 45% and 79% respectively. In addition, the IEP's assessment is that it was potentially 'easy' to fully localise the solar PV value chain, while only 3% of the wind value chain, relating to the turbine (including the gearbox and transformer), was likely to remain imported, unless there was enough global demand to justify the manufacturing investments required to produce them locally. Coal-fired plants also had a relatively high immediate local content of 53%, but producing the balance locally was seen as mostly challenging, with just 10% described as easy to localise at 5 GW. Nuclear is described in the IEP as being the most difficult to localise, with only 36% within domestic industry's immediate grasp and 7% seen as an easy local-content opportunity. In fact, the IEP shows that just half of a nuclear plant's content could be produced locally, and only on the back of strong global demand. To quote directly from the IEP:

> From this it is evident that wind, shale gas, coal mining and CSP have the highest proportion (at least 60%) of spend that could be localised with minimal effort for 5 GW capacity installation. Nuclear plants, due to high complexity and not enough volumes being built from

either a country or regional perspective, cannot be economically local-ised. Similarly CCGT plants are constructed centrally in global loca-tions and are exported to other markets as turnkey projects, it is also not easy to economically localise them.[3]

Using the IEP framework, direct jobs can be divided into two categories: those associated with the building of the power station, which are referred to as capital expenditure, or capex jobs, and those needed to actually run the plant, or operational expenditure (opex) jobs. Direct jobs are the employees directly responsible for building or running the power plants, including any primary energy extraction. Supplier jobs, meanwhile, are those jobs associated with the materials and supplies required to support the construction of operation of the project in question (see Figure 7.1). Capex-related supplier jobs are those created in the factories established to supply the various components and materials for the project, such as steel and cement, or a turbine for a coal-fired plant, blades for a wind turbine, or the PV panel for a solar farm (see Figure 7.2).

A country's ability to localise the goods and services required for a pro-ject will determine the extent to which capex-related supplier jobs can be captured domestically. For the IEP analysis, the measurement used across all technologies for both capex and opex jobs is the number of 'job-years' that the total expenditure per plant creates. The IEP used specific refer-ence factories to calculate the total full-time employees required to produce inputs for 5 GW of each of the generation technologies assessed. These reference factories, located domestically and abroad, that produce the goods used to construct a project.

	Direct jobs	Supplier jobs
Capex jobs	Direct jobs related to the construction of the power station • Developers • Planners • Construction worker • Brick layers	Supplier jobs related to the construction of the power station • Turbine manufacturers • Solar PV panel manufacturers • Cement producers • Steel producers
Opex jobs	Direct jobs related to the operation of the power station • Power plant workers • Coal mine workers • Control room operators	Supplier jobs related to the operation of the power station • Service providers for operation of power stations • Service providers for coal-mining operations

Figure 7.1 Structure of the job-creation-potential assessment as per McKinsey study for the IEP

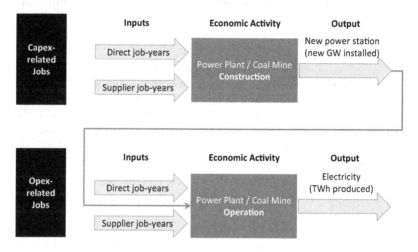

Figure 7.2 Capex-related jobs lead to new power stations, while opex-related jobs, together with an operational power station, leads to electricity produced

Direct jobs

In the following, we will first focus on the direct jobs created during construction and operation of a power station.

The direct employment figure measures capex 'job-years' created for every one new gigawatt (1 GW) installed, while opex job-years are measured against every terawatt-hour (TWh) produced. A job-year basically means that it would take one full-time employee one year to fulfil a certain task, or two full-time employees six months, or four full-time employees three months, and so on. If we know how many job-years are required as an input to produce a certain outcome (in this case a new power station, or a certain amount of electricity from an operational power station), all we need to know in addition is how much of that output we want to build yearly in perpetuity, in order to know how many permanent jobs will be created.

In other words, the number of permanent capex jobs is derived by multiplying the capex job-years that are required to build 1 GW of new capacity of a certain power station technology, with the annual new-build rate of that power station technology in GW per year. For example, if it takes 3 000 full-time employees five years to build 2 GW of new capacity of a certain power station technology, the job-years required to build 1 GW are

7 500 (i.e. 3 000 jobs x 5 years ÷ 2 GW = 7 500 job-years/GW). If 2 GW of that technology are newly commissioned every year in perpetuity, 15 000 permanent jobs are being created. Why? Because in that scenario, in order to commission 2 GW of new capacity every year, at any given point in time, 10 GW would be under construction, because of the build time of five years. If 3 000 full-time employees work on 2 GW, 15 000 full-time employees work on 10 GW. And because these 10 GW are under construction all the time, the 15 000 jobs become permanent.

The number of permanent opex jobs, on the other hand, is calculated by multiplying the opex job-years required to produce 1 TWh of energy from a certain power station type with the amount of annual energy that is produced by that power station type in TWh per year. Once the 2 GW power station described previously has been built and enters operation, the plant will produce a certain annual energy output in TWh per year. For argument's sake, let's assume it is a coal-fired plant producing 10 TWh/a. If it takes 1 000 full-time employees to operate the 2 GW power station and associated coal mine, the job-years required to produce 1 TWh are 100 (1 000 jobs x 1 year ÷ 10 TWh = 100 job-years/TWh). If now 10 TWh are produced from that technology every year in perpetuity, the 1 000 jobs become permanent jobs.

Using this measure and the figures provided in the IEP, the total local-isable capex job-years for solar PV and onshore wind are estimated at 8 724/GW and 6 401/GW respectively. The localisable opex job-years, measured against the plant's energy output, are 107/TWh for solar PV and 120/TWh for wind. By comparison, the capex job-years to construct a coal-fired power station and establish an associated coal mine is a combined 25 571/GW, with the majority of the jobs (24 745) associated with the building of the actual power station. The opex job-years at the coal station, meanwhile, total 77/TWh, with more than half (49/TWh) of those at the mine (see Table 7.1).

Table 7.1 Direct jobs potential associated with different generation technologies[4]

	Unit	Solar PV	Wind	Coal
Capex-related	Job-years per GW (direct)	8 724	6 401	25 571 *Power station: 24 745* *Coal mine: 826*
Opex-related	Job-years per TWh (direct)	107	120	77 *Power station: 28* *Coal mine: 49*

On the face of it, therefore, coal wins the jobs contest hands down. However, to produce 100 TWh a year of electricity from a mix comprising 50% solar PV and 50% wind, which have far lower capacity factors when compared with a traditional baseload plant, there would be a requirement for 25 GW of installed solar PV capacity (which would produce 50 TWh/a) and 20 GW of installed wind capacity (which would produce the other 50 TWh/a). With a life-of-plant assumption of 25 years for solar PV and 20 years for wind, a country would need to build 1 GW a year both of solar PV and wind in perpetuity to meet the 100 TWh-a-year energy requirement – naturally, the replenishment of the solar PV and wind fleets would take place only in the absence of the emergence of a more efficient disruptive technology. By comparison, to produce the same 100 TWh a year from a new-build coal plant, only 15 GW of capacity would be required. In addition, the plant would have a lifetime of 30 years, which translates to a build requirement of roughly 0.5 GW a year in perpetuity.

Taking into account these fundamental differences between variable renewable-energy plants and coal-fired power stations, it is now possible to make fair job-year comparisons for a 100 TWh-a-year energy requirement (see Figure 7.3). Using the numbers generated in the IEP (8 724 capex job-years for every 1 GW of new solar PV installed and 6 401/GW for wind, as well as opex solar PV and wind job-years of 107/TWh and 120/TWh respectively), the capex jobs in the wind and solar PV case are 15 125 and the opex jobs are 11 350, yielding a combined total of 26 475 jobs to produce 100 TWh/a from a combined solar PV and wind fleet. These jobs are now permanent jobs in perpetuity, because there will be constant rebuilding and operations of the 25 GW solar PV and 20 GW wind fleet.

These 26 475 permanent direct jobs are almost 30% higher than the new-coal scenario, where the combined capex and opex job-years are only 20 486 to produce the same amount 100 TWh/a (including coal-mining jobs).

Supplier jobs

If we assume that all jobs which as per the IEP analysis do not require a global demand of South African–made power plants and power plant components could in the long run be localised, we have to add the following supplier jobs to the picture.

On the capex side, 4 917 and approx. 4 500 job-years per new GW are required as input for solar PV and wind respectively in the supplier industry (solar PV module and wind turbine manufacturers). As for opex, 18 and 4 job-years per TWh produced are needed for solar PV and wind. On the coal side, roughly 5 000 plus 993 job-years per new GW are required in the supplier industry for the construction of the power station and the associated

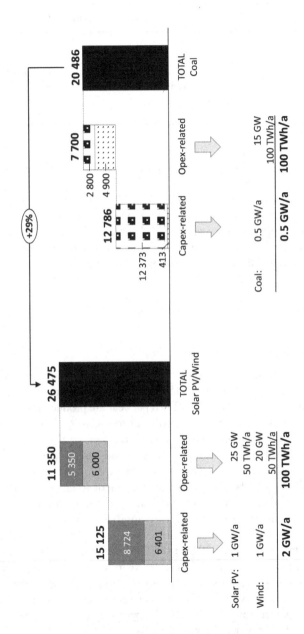

Figure 7.3 Total permanent jobs directly created from producing 100 TWh/a in perpetuity from either a mix of solar PV and wind or from coal-fired power stations

Table 7.2 Supplier jobs potential associated with different generation technologies[5]

	Unit	Solar PV	Wind	Coal
Capex-related	Job-years per GW (supplier)	4 917	4 500	5 993 *Power station: 5 000* *Coal mine: 993*
Opex-related	Job-years per TWh (supplier)	18	4	47 *Power station: 34* *Coal mine: 13*

coal mine, while 34 plus 13 job-years are required from suppliers during operations (see Table 7.2).

Hence, to produce the assumed 100 TWh/a in perpetuity from solar PV and wind requires a total of approx. 37 000 permanent jobs (including a localised supplier industry with 10 527), while to produce the same amount of electricity from coal requires only 28 200 permanent jobs (including 7 697 in the localised supplier industry) (see Figure 7.4).

This means the steady state of continues rebuilding of a 100-TWh-per-year electricity system in the solar PV and wind case creates more than 30% more jobs than a coal-based system.

Job transition

This comparison applies to a world in which the steady state of 25 GW solar PV and 20 GW wind or a 15 GW coal fleet has already been reached, is continuously rebuilt and 100 TWh per year are produced from either the solar PV/wind fleet or from the coal fleet respectively. In reality of course, one needs to transition from today's job numbers in the incumbent coal-fired power industry to the solar PV- and wind-based industry.

If we assume that this transition would happen over a time period that would allow the solar PV and wind deployment to gradually ramp up and displace new-build coal and then electricity produced from coal, we would move from one steady state (today) to another (2050).

Steady state today (pre-2017): 0.5 GW per year of new deployment of coal-fired power stations, 0.5 GW per year decommissioning of old coal-fired power stations (after 30 years of operations), 15 GW constant operational coal capacity and 100 TWh per year of electricity produced from coal-fired power stations. The 15 GW operational coal capacity is located in approx. five individual power stations.

Steady state in future (2050): 1.0 GW of solar PV and wind each would be deployed and decommissioned each year, 25 GW and 20 GW of solar

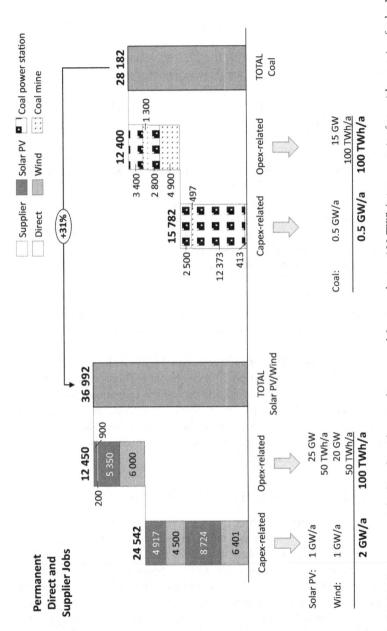

Figure 7.4 Total permanent jobs directly and at suppliers created from producing 100 TWh/a in perpetuity from either a mix of solar PV and wind or from coal-fired power stations

PV and wind were operational in perpetuity and would also produce 100 TWh per year of electricity (50 from solar PV, and 50 from wind). The 25 GW of solar PV are located in hundreds of thousands of individual solar PV power stations across the country (small to large rooftop systems and large, utility-scale power stations), while the 20 GW of wind capacity are located in approx. 400–500 individual wind farms spread across the country.

This transition of new-build, decommissioning and operational asset base is depicted in Figure 7.5.

In today's steady state, to produce 100 TWh per year from coal-fired power stations, 28 200 permanent jobs are required to construct and operate the power stations and the associated coal mines (see previous Figure 7.4). In the future steady state, 37 000 permanent jobs are required to construct and operate the solar PV and wind farms. The transition from the one to the other steady state in terms of job numbers is shown in Figure 7.6, where it is assumed that the supplier jobs for solar PV and wind (manufacturing of solar panels and wind turbines) would only gradually happen after the start of the renewables new-build programme and would come to its full capacity only after 10 years of initial deployment.

Still, because solar PV and wind are more job intensive along the entire value chain than coal-fired power stations, total permanent job numbers would immediately increase as the renewables new-build programme kicks in (here assumed in 2018).

In this comparison, it is not even considered yet that the decommissioning of old coal-fired power stations also creates jobs. Those are not included in the numbers shown and would add during the transition phase to the total number of jobs in the electricity sector. It also does not include the jobs that are created in the building of the necessary flexibility options to accompany the solar PV and wind mix, namely the construction and operations of gas-fired power stations.

Comparison of hypothetical to real world

In the real world, the performance of the renewables industry has been even more impressive than assumed in the IEP. The job-years reported since the start of South Africa's renewables programme in 2011 (which are mainly direct jobs, because a significant supplier industry did not have enough time and investment certainty to establish itself yet) show that, to build the first 3.1 GW of new capacity, 28 152 job-years were created during construction.[6] That's equivalent to 9 081 job-years for every new 1 GW installed, compared with the assumption of 8 724 capex job-years for 1 GW of solar PV, or 6 401 for wind. Meanwhile, as of March 2017, a total of 3 055 job-years had been created from solar PV and wind operations

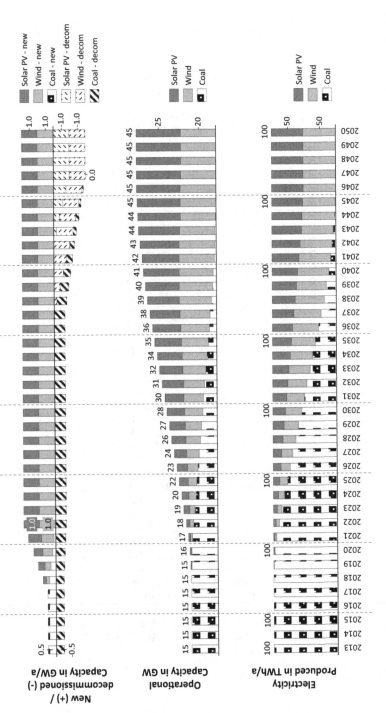

Figure 7.5 Hypothetical structure of supplying 100 TWh/a from coal (today) to supplying the same amount of electricity from solar PV and wind (2050), transitioning from 2013 to 2050

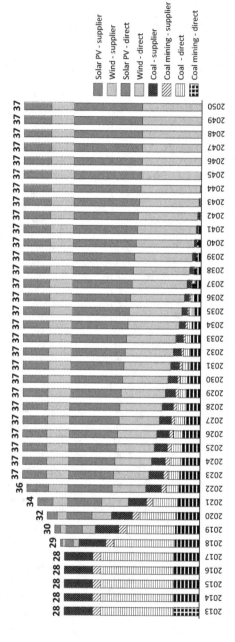

Permanent Jobs ('000)

Figure 7.6 Hypothetical jobs transition from coal to solar PV and wind between 2017 and 2050, assuming a 100 TWh/a electricity production from coal being gradually replaced with 100 TWh/a from solar PV/wind

Legend:
- Solar PV - supplier
- Wind - supplier
- Solar PV - direct
- Wind - direct
- Coal - supplier
- Coal mining - supplier
- Coal - direct
- Coal mining - direct

that produced 15.22 TWh of electricity. In other words, 201 job-years for every 1 TWh produced, which is, once again, higher than the forecasts of 107 for solar PV and 120 for wind. In reality, therefore, jobs associated with South Africa's renewables industry are significantly higher than the initial IEP forecasts, making the actual picture even more attractive than the alternatives.

International experience amplifies the argument that renewables are, in fact, more labour intensive than fossil power generation. In the US, for instance, the number of solar and wind jobs now far exceed those in the coal sector. A comprehensive report[7] on job numbers in the electricity sector produced by the US Department of Energy puts the combined number of solar and wind jobs across the US at 475 000, compared with 160 000 for both coal-fired power generation and coal mining. Total electricity generation from wind and solar in the US in 2016 was 6.5%, while that from coal was 30.4%. Therefore, relative to the electricity generated, the job numbers from wind and solar are even greater than the absolute numbers presented. What's more, solar and wind are creating new jobs 12 times faster than the rest of the US economy. Meanwhile, the International Renewable Energy Agency's 2017 annual review[8] states that the renewable-energy sector employed 9.8 million people, directly and indirectly, in 2016. In addition, renewables-related employment worldwide has grown consistently since the agency's first annual assessment in 2012, when the figure stood at 7.1-million.

Job effects of a least-cost power system expansion

As has been shown in Chapter 3, a least-cost expansion of the South African power system is one that leads to more than 80% energy share in the electricity sector being provided by solar PV and wind. Least cost in itself is a desirable target to aim for, because competitive prices for electricity is at the basis of a competitive economy, especially if it is an energy-intensive one like South Africa's with its mining and heavy industry sectors. But what about the jobs that the electricity sector produces in itself? As we have seen in the previous part of this chapter, wind and solar PV are more labour intensive than coal. Hence a transition from coal to renewables is naturally beneficial for the job market, too. In real numbers, using the least-cost expansion of the power system from Chapter 3, this means that today's permanent jobs that are related with coal-fired power generation in South Africa (coal mining for Eskom and Eskom employees in its 15 coal-fired power stations) of 49 000 would increase to 108 000 in the solar PV, wind and the residual coal sector by 2050. The need to build gas-fired power stations and other complementary flexibility options would add another

10 000 permanent jobs to the new mix. Hence, an increase of the electricity demand by 60% from 243 TWh/a in 2016 to 388 TWh/a in 2050 would be associated with a roughly 150% increase in the number of jobs in the electricity sector – because solar PV and wind are more labour intensive than to produce electricity from coal.

What is clear, therefore, is that there are direct employment benefits of a transition from coal to solar PV and onshore wind. In addition, these are likely to be more resilient than coal jobs that could become increasingly vulnerable in a world that is now firmly on a decarbonisation path. What's more, significant potential exists to create capex and opex supplier jobs around the construction, operation and maintenance of a large-scale renewables fleet. The IEP clearly shows that South Africa's potential to localise the solar PV and wind components used during construction is higher than it is for both coal and nuclear. In fact, the document shows that all of the capex and opex jobs for solar PV are easily 'localisable', while the bulk of the capex and opex supplier jobs for onshore wind could also be localised. By contrast, more than half of the capex and opex supplier jobs associated with nuclear could be localised only if there was 'significant investment' or where there was material 'global demand' for South African-made nuclear components. For coal, all of the opex supplier jobs are seen as easily localisable, but less than half of the capex supplier jobs.

What about the workers?

There is naturally concern about prospects for existing power station employees, as well as the country's coal miners, should South Africa move to replace its coal fleet with one based on solar PV and wind. The net effect over the long term is jobs positive, but that is of little or no consolation for individuals whose livelihoods are currently tied to the energy mineral. This is also not a uniquely South African concern. Many other economies, most notably the US, are experiencing similar anxieties as the future of coal mining comes into question. Not since the showdown between Prime Minister Margaret Thatcher and British coal miners during the 1980s, has coal been as politically emotive. For South Africa's transition to succeed, therefore, a credible, well-communicated and expertly executed plan will be required. The overall objective should be to cushion those most vulnerable, without sacrificing the main goal of migrating to a least-cost, decarbonised and most-jobs electricity future. It goes without saying that such an aspiration is easier articulated than actualised. Indeed, in the absence of a believable strategy that is diligently implemented, the risk of ghost towns and unemployed mineworkers is all too real. So is the threat of a political backlash, notwithstanding the socioeconomic windfalls that are expected to accompany any transition away from coal. Chamber of Mines figures show that, excluding Sasol, the coal-mining sector employs in the region of 80 000 people, the third largest group in the mining sector after gold and platinum group metals. Their annual earnings are in the region of R20 billion.[9]

Making the task somewhat easier is the fact that the total number of jobs will grow, not shrink, should South Africa transition from coal to renewables. Making it politically complex, though, is that these new jobs will be more dispersed. Therefore, the prospect of political mobilisation around renewables jobs is lower when compared with a coal-fired power station where the jobs and the plant and the mine are both geographically concentrated and highly visible. Ensuring that the transition is both just and well managed will likely require a social compact between government and civil society that clearly articulates the benefits of moving away from a sunset industry to one where the growth prospects are strong. The compact should also include a transparent and practical roadmap for how society will navigate the hazards. A key message is that this will be a gradual 20-year process, not an overnight shock.

This is a process that is already under way globally. Coal is thus vulnerable, regardless of South Africa's policy stance. In fact, several countries, or regions within countries, have set firm targets for the closure of their coal-fired power stations. Finland recently decided to phase out all coal-fired power stations by 2030, and so did the province of Alberta in Canada. Alberta's move is most remarkable, as it still produced 75% of all its electricity from coal in 2005 and 50% in 2017. For both jurisdictions, the firm decision to phase out coal, over a reasonably elongated time frame (longer than the typical investment cycle of large refurbishments in existing assets), gives the long-term investment certainty that is most crucial for an orderly transition.

Several commercial banks and development finance institutions have even placed outright bans on funding new coal mines, while others are increasingly reticent to fund power stations. In 2016, world coal production fell by 6.2%, or 231 million tonnes of oil equivalent (mtoe), which the BP Statistical Review of World Energy 2017 described as the largest decline on record. China's production fell by 7.9% or 140 mtoe, also a record decline. US production fell by 19%, or 85 mtoe.[10] The publication also noted that coal production and consumption in the UK fell back to levels last seen almost 200 years ago, around the time of the Industrial Revolution, with the UK power sector recording its first ever coal-free day on 21 April 2017. In addition, coal-mining jobs have become increasingly vulnerable to automation. In fact, research by the International Renewable Energy Agency shows that historically job losses in coal mining were mainly not attributable to the rise of renewables – they happened 10–20 years before renewables were even introduced in a large scale. The 2017 study shows that rising automation in extraction are resulting in job losses in the coal sector in a number of countries, including China and India.[11]

One implication of these global developments is that the market for South African export coal will also decline over time, which means that the appetite among investors to build new domestic mines will diminish. This has repercussions for the cost of coal domestically, as the export component has traditionally made such investments commercially attractive and simultaneously lowered the cost of coal supplied to Eskom power stations. The roadmap associated with the renewables transition cannot ignore these international decarbonisation trends. Instead, it must recognise and integrate them. This has immediate implications for the education and training system, as it would be unwise to train a new generation of coal miners, with limited job prospects. By actively discouraging young entrants into the coal sector, there will be greater alignment between the closure of the power stations and mines and natural-attrition rates. The transition can be engineered such that the number of people employed in the industry declines in line with the regular retirement of workers and with the scheduled retirement of power stations and the associated mines. Simultaneously, the further education and training system should be immediately calibrated to the technical needs of the emerging renewables-based electricity system.

Trade unions should, at the same time, broaden their organisational horizons to include all energy workers, including those involved in building, operating and maintaining solar and wind farms. In this way, their membership could be replenished without triggering a fight-to-the-death campaign to save coal jobs. Labour unions could even lobby for the inclusion of a coal-transition levy or equity as part of the procurement of any new renewable-energy project. Already the Renewable Energy Independent Power Producer Procurement Programme includes socioeconomic and enterprise development as a bidding requirement, along with a stipulation that near-project communities receive equity, through trusts. Therefore, adding a few cents to the tariff, specifically to support coal miners and coal-mining town, should be feasible. Levy funds could be specifically ring-fenced to capitalise re-training schemes, provide bursaries for the children of coal miners, or incentivise the development of alternative economic activities in mining towns. Should equity be the preferred instrument, miner pension funds could receive stakes in the solar and wind projects so that they become direct beneficiaries of the transition to renewables. So, as the one industry declines, the other industry grows and a portion of the equity in the growing industry is given to coal miners through their pensions. Unlike the current debate about whether community trusts should receive benefits earlier in the operational life of a plant, a delayed windfall could work well for mineworkers, where payments could be gleaned at the end of their careers.

Progressive leadership at a local government level could actively encourage the conversion of mining land into solar and wind farms, so as to avoid a cliff-edge scenario associated with closure. Many of these mines are also located close to established electricity network infrastructure, which could facilitate cost-effective grid connection. Other benefits of converting abandoned coal mines into combined wind and solar sites would be the mitigation of some of the local job losses, while assisting with the funding of environmental rehabilitation.

Economic transformation

A similar approach should be adopted with regards to the racial transformation of ownership and control in the electricity industry. Here again, a social compact is essential, particularly given the strides made over the past 20 years in encouraging black economic empowerment in coal mining and logistics. As with workers, entrepreneurs and coal investors are vulnerable to global developments. As the demand for thermal coal comes under pressure internationally, so will the South African industry, whose fortunes are inextricably linked to the global demand for coal. The International Energy Agency forecasts that between today and 2022 growth in renewable generation will be twice as large as that of gas and coal combined. While coal remains the largest source of electricity generation in 2022, renewables are closing the gap.[12] The forecast demonstrates two things: firstly, that the renewables trend is gaining momentum, while coal is weakening; secondly, that there is still time for investors, countries and unions to respond to this shift. For South Africa, which is aspiring to both industrialise and transform the skewed racial composition of its economy, this trend presents risks as well as genuine opportunities.

Should the country decide to ignore the global trends and continue to focus its transformation efforts in the coal sector, it will eventually do so on the back of subsidies, which will be needed to sustain uncompetitive coal-fired power stations. In addition, the potential to industrialise on the back of coal will increasingly be limited to the domestic economy. The net effect could be to lock black South Africans into an unsustainable sunset industry, further cementing economic disadvantages and disparities.

By contrast, the very real possibility now exists for building a fully transformed renewables industry from the bottom-up. Given the scale of the wind and solar PV opportunity, there is room for both large and small black-owned independent power producers (IPPs). The large utility-scale

IPPs will probably initially need partnerships with established global participants. However, the rooftop opportunity is such that there is room for the entry of far smaller firms into the market. Even at the micro-enterprise level, there may be an opportunity. By incentivising solar and wind installations in rural villages and allowing those communities to sell surplus energy back into the grid, the income of subsistence or small-scale commercial farmers could be significantly improved and decoupled from the vagaries of weather- and market-related events. At the corporate level, the added transformational advantage is that the new IPPs would not be established as a result of the sale of existing assets. Instead, an entirely new industry could be created, which is fully transformed from the start.

Likewise, there is an opportunity to incentivise black industrialists to align their manufacturing investments to the needs of solar and wind industries. Here again, partnerships with established technology providers will be important to begin with. However, over time, South Africa could evolve into a manufacturing hub for renewables, supplying not only the domestic market, but also the markets that are likely to take shape across the rest of Africa and the rest of the world. Here it is important to recall the IEP assessment, indicating that it is easier to localise components associated with solar PV and onshore wind than is the case for coal and, even more so, for nuclear. In fact, the IEP shows that more than half of the components associated with the construction of a nuclear power station could only be produced locally in the presence of significant levels of investment and/or global demand for such components.

The domestic construction industry could also be a major beneficiary of the transition, as could the construction materials industry, particularly concrete and steel producers. As with the analysis of job-years, the demand for steel and concrete of an individual wind farm may seem paltry when initially compared with a nuclear reactor, or a mega coal project. However, the picture changes dramatically when aggregated across the entire renewables fleet. To stay with the example of 100 TWh per year electricity production, in order to build a coal fleet of 15 GW that would be able to supply that amount of annual electricity, it would require 7–18 million tonnes of concrete and 2–22 million tonnes of iron and steel. To build a nuclear fleet of 13 GW, which would also produce 100 TWh per year, 2–7 million tonnes of concrete and below 1 million tonne of steel are required. For the electricity-output-equivalent fleet of 25 GW solar PV and 20 GW wind, 20–25 million tonnes of concrete and 4–30 million tonnes of steel would be required.[13]

These numbers have very wide ranges, because the type of power generator can be very different in design within a certain technology class. However, it is clear that, again, solar PV and wind produce the most economic activity, requiring significantly more concrete and steel than either a coal or a nuclear fleet.

Growth platform

However, the most significant benefit lies in what this transition could mean for South Africa's energy competitiveness. The fact that South Africa has better solar and wind resources than just about any other country means that its power will be comparatively cheaper. The upshot is that South Africa could reposition itself as the investment destination of choice for any activity that is electricity intensive.

It may be tempting to dismiss this proposition as a 'back to the future' aspiration that has already been tried but failed, when the country's electricity crisis showed up the shortcomings of a strategy of supporting industry through artificially low power tariffs. Indeed, as South Africa's electricity prices have increased in recent years, the traditional energy-intensive industries of steelmaking, ferrochrome and aluminium smelting, as well as deep-level mining have come under intense competitiveness pressure. However, this strain has arisen not because the logic of deploying low-cost energy as a tool for industrial development and export competitiveness is flawed. Instead, it is because electricity can no longer be kept artificially cheap through tariffs that do not reflect the true cost of production. Nevertheless, in a decarbonising world, South Africa has the opportunity to reclaim its low-cost crown. By virtue of its unrivalled solar, wind and land resources, the country has a genuine competitive advantage of being able to produce clean, cheap electricity from sources that are inexhaustible.

Once this realisation takes hold, the focus of policymakers should then turn to building the economy – especially sectors that require large amounts of energy – based on the country's energy advantage. In that way, South Africa could again become a global anchor point for the beneficiation of those commodities that rely on cheap and abundant electricity. This time round, though, there is the added benefit of producing decarbonised goods, making them immune to any potential move by countries to restrict or penalise trade in carbon-intensive products.

One of these opportunities was flagged in Chapter 4, where the potential for turning South Africa into the 'Saudi Arabia' of green fuels was raised. The idea is to progressively couple the low-cost renewable fleet with the platform that already exists in South Africa, to manufacture liquid fuels and chemicals from coal and natural gas. It may be possible to repurpose Sasol to produce decarbonised fuels by transitioning it from a coal- and gas-to-liquids to a power-to-liquids platform. In this way, South Africa could become a global supplier of a decarbonised jet fuel in the non-electrifiable aviation sector, which may well even attract a premium.

Furthermore, there is potential to deploy South Africa's renewables-led electricity mix to produce hydrogen and hydrogen derivatives for export and to fuel the shipping industry. It may even prove feasible to manufacture 'green fertilisers' by combining hydrogen produced using renewables-fuelled electrolysers with nitrogen to manufacture ammonia. Cheap electricity from the renewables fleet could also power desalination plants, creating the water resources needed not only to sustain households in certain cities, but also to offer security of water supply needed to expand agricultural production and exports.

Sector coupling is also likely to improve visibility of other economic spinoffs and benefits, not least the opportunity to lower the cost of transport and to materially improve the country's trade balance, through supporting the adoption of electric vehicles (EVs). In fact, the initial signs are that these benefits could be captured almost immediately, both for the car owner and the country (refer to Chapter 4).

Bottom Line: Overall, the 'convenient truth' is that South Africa is in a strong position to decarbonise its energy mix cost effectively and without undermining security of supply, jobs or the economy. In fact, this decarbonised platform will be cheaper than any other mix currently being contemplated. Because South Africa has better combined solar and wind resources than just about any other country, its power will be comparatively cheaper. The upshot is that South Africa can reposition itself as the investment destination of choice for any activity that is electricity intensive. Building and operating an electricity system based on solar, wind and flexible generation technologies will create more jobs than any of the alternatives. South Africa is extremely well positioned to pursue an 'electrification-of-almost-everything' future, where the decarbonised electricity system powers a competitive industrial economy, drives an electric-mobility revolution and creates new export and investment opportunities. A convenient truth indeed!

Notes

1 Department of Energy. *Integrated Energy Plan Annexure B: Macroeconomic Assumptions*, 2016.
2 Department of Energy. *Integrated Energy Plan Annexure B: Macroeconomic Assumptions*, 2016.
3 South African Department of Energy, Integrated Energy Plan, Annexure B: Macroeconomic Assumptions, page 27, http://www.energy.gov.za/files/IEP/2016/IEP-AnnexureB-Macroeconomic-Assumptions.pdf.
4 Department of Energy. *Integrated Energy Plan Annexure B: Macroeconomic Assumptions*, 2016.
5 Department of Energy. *Integrated Energy Plan Annexure B: Macroeconomic Assumptions*, 2016.
6 RSA Department of Energy. *National Treasury and the Development Bank of Southern Africa*. Independent Power Producers Procurement Programme, Overview As at 31 March 2017, March 2017.

7 US Department of Energy. *US Energy and Employment Report*, January 2017.
8 IRENA. *Renewable Energy and Jobs*, Annual Review 2017.
9 Chamber of Mines website. *Coal Mining in South Africa*, 2017.
10 BP. *BP Statistical Review of World Energy 2017*, June 2017.
11 IRENA. *Renewable Energy and Jobs*, Annual Review 2017.
12 International Energy Agency. *Renewables 2017*, 4 October 2017.
13 Ashby, Mike, Attwood, Julia. *Fred Lord: Materials for Low-Carbon Power – a White Paper.*

Index

Note: Page numbers in *italic* indicate a figure and page numbers in **bold** indicate a table on the corresponding page